Controlled Branching Processes

To Pilar and our daughter, Pilar,
and sons, Miguel and Manuel

To Raúl and our sons, Jorge and Marcos,
and daughter Sara

To Reneta and Petar

Branching Processes, Branching Random Walks and Branching Particle Fields Set

coordinated by
Elena Yarovaya

Volume 2

Controlled Branching Processes

Miguel González Velasco
Inés M. del Puerto García
George P. Yanev

iSTE

WILEY

First published 2018 in Great Britain and the United States by ISTE Ltd and John Wiley & Sons, Inc.

ISTE Ltd
27-37 St George's Road
London SW19 4EU
UK

www.iste.co.uk

John Wiley & Sons, Inc.
111 River Street
Hoboken, NJ 07030
USA

www.wiley.com

Library of Congress Control Number: 2017956887

British Library Cataloguing-in-Publication Data
A CIP record for this book is available from the British Library
ISBN 978-1-78630-253-3

MIX
Paper from responsible sources
FSC® C013604

Contents

Foreword

Since the 1940s, when Kolmogorov introduced the acclaimed term "branching", the theory of branching processes has been developing in a "supercritical" fashion, spreading in various directions. Nowadays, branching processes have a prominent role in stochastic modeling with their important applications. From nuclear reactions and cosmic rays to cell proliferation and digital information, branching models are used in explaining very interesting real-world stochastic phenomena.

In the beginning, classical branching processes with independent individual evolutions were investigated. Traditionally, they were interpreted as evolutionary models for isolated populations that develop using only their own resources. For the most part, the mathematical theory explored the method of probability-generating functions in obtaining intricate results. The fundamental investigations were summarized in several monographs, appropriately included in the references of the present book.

However, in many real phenomena, the evolutions of the individuals (particles) are not independent. Various approaches to modeling dependent evolutions exist in the literature. The subject matter of this monograph focuses on one of them in detail – random control. In this first book on "controlled branching processes" (CB processes), the authors study the most well-developed class of CB processes having single-type individuals and discrete-state space. The CB processes can be interpreted as models of population dynamics taking into account outside (environmental) factors. For instance, a random part of one generation could be eliminated and would not take part in further evolution (emigration) and/or new individuals could join the population and participate in the reproduction process (immigration). It is interesting to point out that, as a result of the random control (or regulation) mechanism, new effects arise in the population dynamics which were not observed in the classical theory.

The book does not require any previous knowledge of branching processes. From the first chapter, the reader will be introduced to the basic ideas and results of the classical theory. Undergraduate level stochastic education will be enough to follow the exposition. The book can be used in specialized courses on branching processes or as a supplementary text for a general course on applied stochastic processes. The monograph will be also very useful for researchers who would like to apply some form of control (regulation) in modeling concrete population dynamics. The specialists in branching processes will also be surprised to find some interesting ideas and open problems. For instance, Chapter 5 discusses some recent statistical investigations on CB processes. It seems that one of the main directions for further investigations is the development of the theory of time-continuous CB processes.

In conclusion, let me commend the authors for the easy-to-follow mathematical exposition accompanied with appropriate discussions and illustrations.

Nikolay M. YANEV
Sofia, Bulgaria

Preface

"All theory is gray, my friend. But forever green is the tree of life".
– Johann Wolfgang von Goethe, Faust: First Part

The intent of this book is to provide stimulating discussion and insights into the available results for discrete-time branching processes with random control functions. These stochastic models of population dynamics have evolved from the classical branching processes as models which assume a state-dependent reproduction of the population. A fundamental assumption in traditional branching processes is that each individual's reproduction or survival is independent of the chance of others. As a result, branching processes are good models of the evolution of small populations, in which resource limitations, for example, do not play an important role. Naturally, we are also interested in processes in which generation size increases or decreases depending on the available resources or interactions with other populations.

Since the 1960s, a number of models allowing different forms of population size regulations have been introduced and studied. In 1974, Sevastyanov and Zubkov [SEV 74] proposed a class of branching processes in which the number of reproductive individuals in one generation decreases or increases depending on the size of the previous generation through a set of control functions. The individual reproduction law (offspring distribution) is not affected by the control and remains independent of the population size. These processes are known as controlled branching processes (CB processes). In 1975, Yanev [YAN 75] (no relation to the third author of this book) essentially extended the class of CB processes by considering random control functions. CB processes constitute a very large class of stochastic processes, which take into account different conditions for immigration and emigration. These have been successfully used as modeling tools in a wide range of applications outside of mathematics. Within mathematics, CB processes are a fascinating research field of their own with thought-provoking unanswered questions. Over the years, a number of particular subclasses of CB processes have been introduced and investigated in detail. At the same time, fruitful connections were established to other types of branching processes, including two-sex processes and population-size-dependent processes. We consider the general properties of CB processes rather than their applicability to any real-world system. In particular, we turn our attention to (i) the probability of extinction, (ii) criticality, (iii) limit theorems, and (iv) statistical inferences.

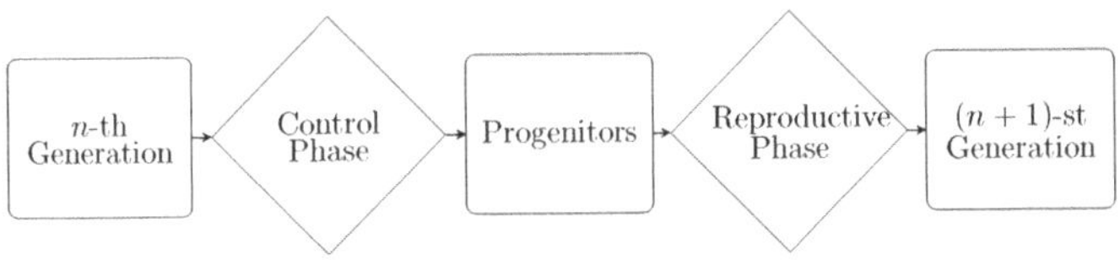

CB processes are discrete-time and discrete-state stochastic population models. The two qualifiers, discrete and stochastic, simultaneously provide richness and technical challenges in terms of the measurements that can be made. Population development is modeled in two phases: the *reproductive phase*, when the individuals' produce offspring, and the *control phase*, when the number of potential progenitors is determined (see the diagram above).

This book is divided into three distinct parts. The first part, consisting of two chapters, discusses particular sub-classes of CB processes of varying generality. Chapter 1 is devoted to classical models, including Bienaymé–Galton–Watson processes, processes that allow for unrestricted immigration, processes with immigration at zero only, and processes with time-dependent immigration. Chapter 2 presents in detail one class of processes with migration (immigration and emigration). An extension, connecting processes with migration and alternating regenerative processes, is also discussed. The second part includes Chapters 3 and 4, in which CB processes are treated in their generality. Chapter 3 addresses the fundamental problems of extinction and classification into subcritical, critical, and supercritical subclasses. Chapter 4 presents a variety of limit theorems depending on the criticality of the processes. The third part, consisting of Chapter 5, addresses statistical estimation procedures for certain parameters of CB processes. Each chapter ends with some background and bibliographical notes. For easy reference, some classical and auxiliary results needed in the proofs are given in the appendices.

The authors express their appreciation to N.M. Yanev (who is the third author's academic advisor) for his continual support and mentorship. The first two authors would like to express their gratitude to the members of the research group *Branching Processes and their Applications* at the University of Extremadura (Badajoz, Spain), especially to M. Molina, R. Martínez and C. Minuesa for their contributions to the development of the theory of CB processes. The third author thanks his colleagues and teachers I. Rahimov and J. Stoyanov for their support and helpful discussions.

Part of the research included in this book was supported by the Spanish Ministry of Economy and Competitiveness (Ministerio de Economía y Competitividad) through the Grant MTM2015-70522-P and by the NFSR 190 at the MES of Bulgaria, Grant No DFNI-I02/17. The research of the first two authors has also been partially supported by the Junta de Extremadura, Grant IB16099, and the *Fondo Europeo de Desarrollo Regional* (FEDER). Working on this book, the third author was on leave from the Institute of Mathematics and Informatics at the Bulgarian Academy of Sciences.

We extend our sincere thanks to E. Yarovaya, the editor of the series of which this book is a part. We also express gratitude to the ISTE editorial office for all the services they provided.

Miguel GONZÁLEZ
Inés M. DEL PUERTO
George P. YANEV
October 2017

1 Classical Branching Models

"Common sense is not so common".
– Voltaire

In this chapter, some classical theorems for the ordinary Bienaymé–Galton–Watson process and its modifications allowing immigration are discussed. These are well-studied models for which a comprehensive mathematical theory exists. It is not our goal to present the most general models and results available, but rather to discuss selected processes with and without immigration, relevant to the material included in the following chapters. Questions of interest for each model include (1) extinction criteria and criticality classification, (2) asymptotic behavior of the survival probability, (3) mean population growth rate, (4) limit theorems for the population size and (5) effects on the population of possible immigration.

1.1. Bienaymé–Galton–Watson process

In this section, the simple branching process with discrete time and one type of individuals is described and its basic results are summarized. This classical model is called Galton–Watson (GW) process before Bienaymé's priority was recognized; now it is often called Bienaymé–Galton–Watson process. In the beginning, the fundamental equation for the probability generating functions (p.g.f.s) associated with the process is derived. Next, the first two moments as well as the coefficient of variation are calculated. The processes are classified according to the value of their offspring mean and the extinction probability in each class is studied. Limit theorems for the population size are discussed with either detailed or outlined proofs.

Let $(\Omega, \mathcal{A}, P)$ be a probability space on which an array of non-negative integer-valued random variables $\{X_{n,i} : n = 0, 1, \ldots; i = 1, 2, \ldots\}$ is given, where $\{X_{n,i}\}$ are independent and identically distributed (i.i.d.) with a common probability mass function (p.m.f.) $\{p_k\}_{k \geq 0}$. To avoid trivialities, we assume that $p_0 > 0$, $p_0 + p_1 < 1$ and therefore $p_j \neq 1$ for any j.

DEFINITION 1.1.– *The (Bienaymé–)Galton–Watson branching process is a discrete time homogeneous Markov chain $\{Z_n\}_{n \geq 0}$ defined inductively by $Z_0 = 1$ and for $n = 0, 1, \ldots$*

$$Z_{n+1} = \begin{cases} \sum_{i=1}^{Z_n} X_{n,i} & \text{if } Z_n > 0 \\ 0 & \text{if } Z_n = 0. \end{cases} \qquad [1.1]$$

By convention, equality between random variables stands for equality in distribution. Alternatively, we will also use the notation $\stackrel{d}{=}$.

The most common interpretation of this formal definition is in terms of the evolution of a population. The state space consists of a number of "individuals", where an "individual" might be an animal or a plant, but also a cell or an elementary particle – the defining property is that it gives birth, splits into, or somehow generates new individuals (see [HAC 05, p. 6]). In the beginning, the population consists of one individual (the ancestor), $Z_0 = 1$, with unit lifetime. In the end of its life, the ancestor produces a random number $X_{0,1}$ offspring (direct descendants) with p.m.f. $\{p_k\}_{k \geq 0}$. Every direct descendant (if there is any) also has a unit lifetime and in its end produces (independently of the other descendants) a random number of offspring according to the same "offspring distribution" $\{p_k\}_{k \geq 0}$. Proceeding in this way and interpreting n as the generation index and $X_{n,i}$ as the number of direct offspring of the ith individual from the nth generation, we can say that Z_{n+1} represents the size of the next $(n+1)$st generation, i.e. the GW process counts the generation sizes.

Note that the above-mentioned terms "birth" and "offspring" are used in a general sense. For example, in an epidemic outbreak, the number of infected individuals represents the states of the branching process and a birth means a new infection and offspring means the number of new infections produced. Generally, in a GW process, the birth of offspring implies death of the parent (non-overlapping generations). However, if the parent gives birth to j offspring and survives, then the number of new offspring, counting the surviving parent, is $j + 1$ with corresponding probability p_{j+1}. This latter assumption can be thought of as a death of the parent and replacement by an "identical substitute". Such an assumption is reasonable in an epidemic outbreak, where the infected parent continues to spread infection until death or recovery (see [ALL 15, p. 2]).

An equivalent to [1.1] representation is (see [KIM 15, p. 38]) $Z_0 = 1$ and

$$Z_{n+1} = \begin{cases} \sum_{i=1}^{X_{0,1}} Z_n^{(i)} & \text{if} \quad X_{0,1} > 0 \\ 0 & \text{if} \quad X_{0,1} = 0, \end{cases} \qquad [1.2]$$

where the random variables $Z_n^{(i)}$, $i = 1, 2, \ldots$ are i.i.d. copies of Z_n. The relation [1.2] reflects the fact that the $(n + 1)$st generation is the sum of the nth generations of subprocesses initiated by the first-generation offspring of the ancestor.

Alternatively, the GW process $\{Z_n\}_{n\geq 0}$ is defined as a Markov chain on the non negative integers with stationary transition probabilities given by

$$P(Z_{n+1} = j | Z_n = i) = \begin{cases} p_j^{*i} & \text{if} \quad i \geq 1,\ j \geq 0 \\ \delta_{0j} & \text{if} \quad i = 0,\ j \geq 0, \end{cases}$$

where δ_{ij} is the Kronecker delta and

$$p_j^{*i} := \sum_{j_1+j_2+\ldots+j_i=j} p_{j_1} p_{j_2} \cdots p_{j_i}$$

is the jth term of the ith fold convolution of the sequence $\{p_k\}_{k\geq 0}$.

Yet another characterization of GW process is through the so-called branching (additive) property. Let $\mathbf{N}^+$ be the set of positive integers. If $Z_n(i)$ denotes a GW process initiated by $i \in \mathbf{N}^+$ ancestors, the family of processes $\{Z_n(i), i \in \mathbf{N}^+\}_{n\geq 0}$ is such that

$$Z_n(i+j) = Z_n^{(1)}(i) + Z_n^{(2)}(j), \qquad i,\ j \in \mathbf{N}^+,$$

where $Z_n^{(1)}(\cdot)$ and $Z_n^{(2)}(\cdot)$ are two independent copies of $Z_n(\cdot)$. It can be shown that this property uniquely characterizes GW processes. It is inherited by more general classes of branching processes (e.g. continuous state processes, e.g. [LAM 67]).

1.1.1. *Moments and probability of extinction*

The offspring p.g.f. $f(s) := \sum_{k=0}^{\infty} p_k s^k$ and the generation size p.g.f. $F_n(s) := E\left[s^{Z_n}\right], 0 \leq s \leq 1$, are the main tools in the study of the process. Recall that, in order to exclude trivial situations, we assumed, $p_0 = f(0) > 0$ and $p_0 + p_1 < 1$. This last condition implies that $f(s)$, which is increasing on $[0,1]$, is a strictly convex function. The following fundamental equations hold.

THEOREM 1.1.– *For any* $n = 0, 1, \ldots$

$$F_{n+1}(s) = F_n(f(s)) = f(F_n(s)) = f_{n+1}(s), \qquad F_0(s) = s, \qquad [1.3]$$

where $f_n(s)$ *is the nth functional iterate of the offspring p.g.f.* $f(s)$.

Proof. The initial condition reflects the assumption $Z_0 = 1$. By the definition [1.1] of the process and the independence of the offspring random variables $\{X_{n,i} : n = 0, 1, \ldots;\ i = 1, 2, \ldots\}$, using the "law of total expectation", we have

$$\begin{aligned} F_{n+1}(s) &= E\left[E\left[s^{Z_{n+1}} | Z_n\right]\right] \\ &= E\left[E\left[s^{X_{n,1}+\ldots+X_{n,Z_n}} | Z_n\right]\right] \\ &= E\left[E\left[s^{X_{0,1}}\right]^{Z_n}\right] \\ &= F_n(f(s)). \end{aligned}$$

Similarly, using [1.2] and "first-step analysis" we obtain

$$
\begin{aligned}
F_{n+1}(s) &= E\left[E\left[s^{Z_{n+1}}|Z_1\right]\right] \\
&= E\left[E\left[s^{Z_n^{(1)}+\ldots+Z_n^{(Z_1)}}|Z_1\right]\right] \\
&= f(F_n(s)).
\end{aligned}
$$

Since $F_1(s) = f(s)$, iterating we have $F_n(s) = f_n(s)$. ∎

The **offspring mean** and **variance** are denoted by m and σ^2, respectively. Throughout this chapter, we assume that $m < \infty$, non-explosive case (sample paths do not approach infinity for finite time). Unless stated otherwise, for the offspring variance we assume $\sigma^2 < \infty$. Next, we will calculate the mean and variance of Z_n in terms of m and σ^2.

THEOREM 1.2.– *The mean and variance of Z_n are $E[Z_n] = m^n$ and*

$$
Var[Z_n] = \begin{cases} \sigma^2 n & \textit{if} \quad m = 1 \\ \sigma^2 \dfrac{m^{n-1}(m^n - 1)}{m - 1} & \textit{if} \quad m \neq 1. \end{cases} \qquad [1.4]
$$

Proof. Differentiating each side of [1.3] and evaluating the derivatives at $s = 1$, we have

$$
E[Z_{n+1}] = F'_{n+1}(1) = f'(1)F'_n(f(1)) = mE[Z_n].
$$

Thus, in general, $E[Z_n] = m^n$. Differentiating each side of [1.3] twice and evaluating at $s = 1$, we have

$$
E[Z_n(Z_n - 1)] = F''_{n+1}(1) = (f'(1))^2 F''_n(f(1)) + f''(1)F'_n(f(1)).
$$

Now, it is not difficult to calculate

$$
E[Z_n(Z_n - 1)] = \begin{cases} f''(1)n & \text{if} \quad m = 1 \\ f''(1)\dfrac{m^{n-1}(m^n - 1)}{m - 1} & \text{if} \quad m \neq 1. \end{cases} \qquad [1.5]
$$

Since $Var[Z_n] = E[Z_n(Z_n - 1)] + E[Z_n] - (E[Z_n])^2$, referring to [1.5] and the formula for $E[Z_n]$, we obtain [1.4]. ∎

For the **coefficient of variation** of the process, $CV[Z_n]$ say, we obtain

$$CV[Z_n] := \frac{\sqrt{Var[Z_n]}}{E[Z_n]} = \begin{cases} \rightarrow \dfrac{\sigma}{\sqrt{m(m-1)}} & \text{if} \quad m > 1 \\ \sigma\sqrt{n} & \text{if} \quad m = 1 \\ \sigma m^{-n/2}\sqrt{\dfrac{1-m^n}{m(1-m)}} & \text{if} \quad m < 1, \end{cases} \qquad [1.6]$$

where the limit is as $n \to \infty$.

The asymptotic behavior of $E[Z_n]$ and $Var[Z_n]$ varies depending on the offspring mean m. If $m < 1$, then both mean and variance of the population size decrease to zero, and the process should eventually die out. If $m = 1$, then the mean remains one, whereas the variance experiences a linear growth. In this case, a more delicate analysis of the asymptotic behavior of the process is required. Finally, if $m > 1$, then both mean and variance increase exponentially, such that the coefficient of variation stabilizes and thus the population experiences a steady expansion. Depending on its asymptotic behavior, the GW processes form three distinct classes as follows.

DEFINITION 1.2.– *A GW process with offspring mean m is said to be subcritical if $m < 1$; critical if $m = 1$, or supercritical if $m > 1$.*

This classification is closely related to the **probability of extinction**, i.e. the probability of absorption at 0.

DEFINITION 1.3.– *The event "$Z_n = 0$ for some $n \in \mathbf{N}^+$" is called extinction.*

A random variable of interest is the **extinction time** T, which is defined with: $Z_{T-1} > 0$ and $Z_T = 0$, i.e. the moment $\{Z_n\}$ hits zero (if $Z_n > 0$ for all $n \in \mathbf{N}^+$, then $T = \infty$). Note that we identify generations and time. As we will see in the next theorem, $P(T < \infty) = q$, where q denotes the probability of the extinction event. The following theorem connects the probability of extinction with the criticality of the process.

THEOREM 1.3.– *If* $m \leq 1$, *then* $P(T < \infty) = 1$, *i.e. the extinction is certain. If* $m > 1$, *then* $P(T < \infty) < 1$, *i.e. there is a positive probability for non-extinction. Furthermore, the probability of extinction* q *is the smallest root of the equation* $f(s) = s$ *for* $s \in [0, 1]$.

Proof. Consider the probability

$$P(T \leq n) = P(Z_n = 0) = F_n(0) = f_n(0). \qquad [1.7]$$

Since $Z_n = 0$ implies $Z_{n+1} = 0$, we have $q = \lim_{n\to\infty} P(T \leq n) = P(T < \infty)$. Let us find an equation for q. Since the probabilities $P(T \leq n)$ for $n = 1, 2, \ldots$ in [1.7] form a bounded and monotonic sequence, we obtain for $0 \leq q \leq 1$

$$q = \lim_{n\to\infty} f_n(0) = \lim_{n\to\infty} f(f_{n-1}(0)) = f(\lim_{n\to\infty} f_{n-1}(0)) = f(q).$$

By assumption $f(0) = p_0 > 0$ and since $f''(s) > 0$, the function $f(s)$ is convex. If $f'(1) \leq 1$, then the equation $f(s) = s$ has a unique root 1 in $[0, 1]$. If $f'(1) > 1$, then in addition to 1, there is a unique root q_1 in $[0, 1)$. It is not difficult to prove that $\lim_{n\to\infty} f_n(0) = q_1$ and then $q_1 = q$. ∎

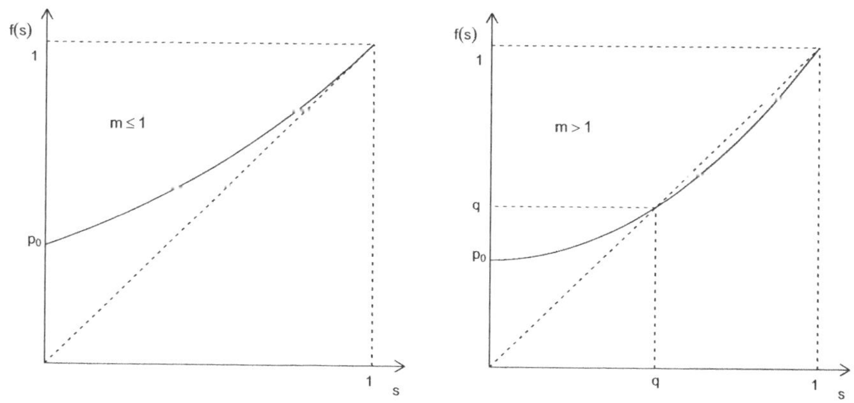

Figure 1.1. *Offspring p.g.f. and extinction probability. Reprinted from [GON 10], with permission from SEIO*

According to theorem 1.3, the extinction occurs with probability 1 not only when $m < 1$, but when the offspring mean is also 1 (excluding the deterministic process $Z_n \equiv 1$). Sometimes this is called **branching paradox**. If $m > 1$, then the probability of extinction is less than 1 (but could be positive), being the unique root of $f(s) = s$ for $s \in [0, 1)$. On the other hand, if $m > 1$, then the family tree of the ancestor is infinite (escaping extinction) with positive probability (see Figures 1.1 and 1.2).

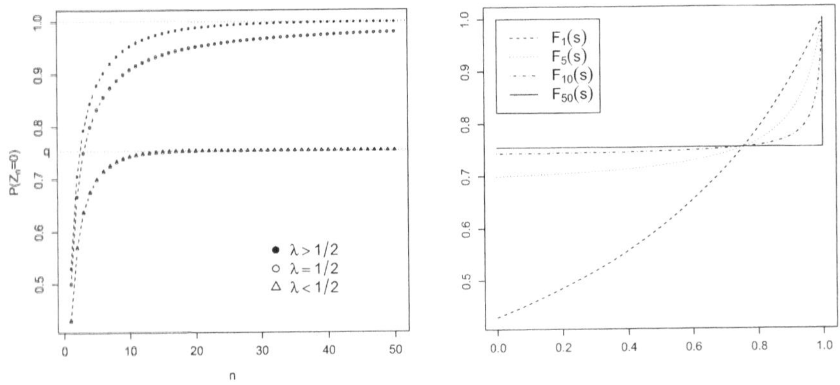

Figure 1.2. *Left: Extinction probabilities for GW processes with geometric offspring distribution with parameter λ, i.e. $p_k = (1-\lambda)^k\lambda$ for $k \geq 0$. In this case, $m = (1-\lambda)/\lambda$. Thus, according to whether $\lambda > 1/2, = 1/2$ or $< 1/2$, the process is subcritical, critical or supercritical, respectively. Right: Evolution of the p.g.f. of Z_n in the supercritical case. Reprinted from [GON 10], with permission from SEIO*

Note that the state 0 is absorbing for the Markov chain $\{Z_n\}_{n\geq 0}$, and it is accessible from each state. It is then easy to deduce that all other states are transient, hence either $Z_n \to 0$ or $Z_n \to \infty$ as $n \to \infty$, i.e. the so-called **extinction–expansion dichotomy** holds.

In the critical and subcritical processes, it is natural to ask about the speed of convergence of $P(Z_n > 0)$ (or equivalently of $P(T > n)$) to zero as $n \to \infty$. It turns out (e.g. [ATH 72, p. 19], [JAG 75, p. 29] and [SLA 68]) that for positive constants c and c', and $\alpha \in (0, 1]$

$$P(Z_n > 0) \sim \begin{cases} cm^n & \text{iff } E[X_{0,1}\log(1+X_{0,1})] < \infty, \text{ when } m < 1 \\ c_\alpha n^{-1/\alpha} & \text{if } f(s) = s + c'(1-s)^{1+\alpha}(1+o(1)),\ s \to 1^-, \end{cases} \quad [1.8]$$

where $c_\alpha = (c'\sigma^2/2)^{-1/\alpha}$. Clearly, the assumption for $f(s)$ in the lower part of [1.8] implies that $m = 1$ and when $\alpha = 1$ and $c_1 = 2/\sigma^2$, it yields $P(Z_n > 0) \sim (2/\sigma^2)n^{-1}$.

1.1.2. *Limit theorems*

The exact form of the distribution of Z_n is not known except in some very special cases. We will see, however, that in spite of the difficulties of calculating the distribution of Z_n, it is possible to obtain detailed information on its asymptotic behavior when the time (generation index) $n \to \infty$. The following equation for the p.g.f. of the process will play an important role throughout this section. Referring to [1.3], we have

$$\begin{aligned} E\left[s^{Z_n}|Z_n > 0\right] &= \sum_{k=0}^{\infty} P(Z_n = k|Z_n > 0)s^k \\ &= \frac{F_n(s) - F_n(0)}{1 - F_n(0)} \\ &= 1 - \frac{1 - F_n(s)}{1 - F_n(0)} \\ &= 1 - \frac{1 - f_n(s)}{1 - f_n(0)}. \end{aligned} \quad [1.9]$$

First, let us consider **the subcritical process**. Since the extinction in this case is certain, the asymptotic behavior of Z_n should be investigated given non-extinction by the nth generation. Theorem 1.2 and [1.8] yield as $n \to \infty$

$$E[Z_n \mid Z_n > 0] = \frac{E[Z_n]}{P(Z_n > 0)} \to c^{-1} < \infty.$$

This suggests that the population size Z_n stabilizes as $n \to \infty$ given that the process has survived by time n. Theorem 1.4 confirms this.

THEOREM 1.4.– *Let $\{Z_n\}_{n\geq 0}$ be a subcritical GW process. Then for $k = 0, 1, \ldots$*

$$\lim_{n\to\infty} P(Z_n = k \mid Z_n > 0) = r_k, \qquad \sum_{k=1}^{\infty} r_k = 1, \qquad [1.10]$$

and $F(s) = \sum_{k=1}^{\infty} r_k s^k$ is the unique p.g.f. solution of the Schröder equation

$$1 - F(f(s)) = m(1 - F(s)), \qquad 0 < s \leq 1; \qquad F(0) = 0. \qquad [1.11]$$

The limit law has a finite mean if and only if $E[X_{0,1} \log(1 + X_{0,1})] < \infty$.

COMMENT.– Theorem 1.4 is a limit theorem of the so-called Yaglom type. Given extinction at some (finite a.s.) future time, a discrete limiting distribution is obtained. The limit $\{r_k\}_{k\geq 1}$ is called a **quasi-stationary distribution**.

Proof. The proof makes use of the continuity theorem for p.g.f.s. Namely, to establish [1.10] it is sufficient to prove the convergence of the generating functions of both sides of [1.10]. Equivalently, we need to prove the convergence of the right-hand side of [1.9]. Clearly, $(1 - f(s))/(1 - s) \uparrow m$ as $s \uparrow 1$. Therefore,

$$\frac{1 - f_{n+1}(s)}{1 - f_n(s)} = \frac{1 - f(f_n(s))}{1 - f_n(s)} \geq \frac{1 - f(f_n(0))}{1 - f_n(0)}$$

or, equivalently,

$$\frac{1 - f_{n+1}(s)}{1 - f_{n+1}(0)} \geq \frac{1 - f_n(s)}{1 - f_n(0)}.$$

Hence,

$$\lim_{n\to\infty} \frac{1 - f_n(s)}{1 - f_n(0)} = 1 - F(s)$$

exists and hence, by [1.9], the limit in [1.10] does too. We proceed with the proof of [1.11].

$$\begin{aligned} 1 - F(s) &= \lim_{n\to\infty} \frac{1 - f_{n+1}(s)}{1 - f_{n+1}(0)} \\ &= \lim_{n\to\infty} \left[\frac{1 - f_n(f(s))}{1 - f_n(0)} \frac{1 - f_n(0)}{1 - f(f_n(0))} \right] \\ &= \lim_{n\to\infty} \frac{1 - f_n(f(s))}{1 - f_n(0)} \lim_{n\to\infty} \frac{1 - f_n(0)}{1 - f(f_n(0))} \\ &= \frac{1}{m}(1 - F(f(s))). \end{aligned}$$

Therefore, $1 - F(f(s)) = m(1 - F(s))$. Furthermore, letting $s \uparrow 1$ we obtain the equation $1 - F(f(1)) = m(1 - F(1))$. Hence, $F(1) = 1$, i.e. the limiting (quasi-stationary) distribution is proper.

Next we establish the uniqueness of the solution. Suppose $G(x)$ is another p.g.f. solution of [1.11]. Differentiating [1.11] we see that both F and G satisfy

$$F'(f_n(s))f_n'(s) = m^n F'(s). \qquad [1.12]$$

For any $s \in [0, 1)$, there is a k such that

$$f_k(0) \leq s \leq f_{k+1}(0),$$

and hence by [1.12]

$$\frac{G'(s)}{F'(s)} = \frac{G'(f_n(s))}{F'(f_n(s))} \leq \frac{G'(f_{n+k+1}(0))}{F'(f_{n+k}(0))} = \frac{G'(f_{n+k+1}(0))}{F'(f_{n+k+1}(0))} \frac{F'(f_{n+k+1}(0))}{F'(f_{n+k}(0))}.$$

From the first equality above, we get

$$\frac{G'(f_n(0))}{F'(f_n(0))} = \frac{G'(0)}{F'(0)}, \qquad n \geq 1,$$

and hence

$$
\begin{aligned}
\frac{G'(s)}{F'(s)} &\le \frac{G'(0)}{F'(0)} \frac{F'(f_{n+k+1}(0))}{F'(f_{n+k}(0))} \\
&= \frac{G'(0)}{F'(0)} \frac{f'_{n+k}(0)}{f'_{n+k+1}(0)} m \qquad \text{(applying [1.12])} \\
&= \frac{G'(0)}{F'(0)} \frac{m}{f'(f_{n+k}(0))}.
\end{aligned}
$$

Letting $n \to \infty$, we get $G'(s)/F'(s) \le G'(0)/F'(0)$. The converse inequality is established similarly. Since $G(0) = F(0)$, this implies the uniqueness.

It remains to prove that the expectation of the limit law is finite if and only if $E[X_{0,1} \log(1 + X_{0,1})] < \infty$. Using [1.11], we obtain

$$
\begin{aligned}
1 - F(f_n(s)) &= 1 - F(f(f_{n-1}(s))) \\
&= m(1 - F(f_{n-1}(s))) \\
&= m^n(1 - F(s)).
\end{aligned}
$$

Therefore, taking into account that $F(0) = 0$, we have

$$
\begin{aligned}
F'(1) &= \lim_{n\to\infty} \frac{1 - F(f_n(0))}{1 - f_n(0)} \\
&= \lim_{n\to\infty} \frac{m^n(1 - F(0))}{P(Z_n > 0)} \\
&= \lim_{n\to\infty} E[Z_n \mid Z_n > 0] \\
&< \infty
\end{aligned}
$$

if and only if $E[X_{0,1} \log(1 + X_{0,1})] < \infty$ by [1.8]. ∎

We turn to **the critical case**. The probabilities $P(Z_n = j \mid Z_n > 0)$ tend to zero for all j. However, by theorem 1.2 and [1.8] with $\alpha = 1$ and $\sigma^2 < \infty$, we have as $n \to \infty$

$$
E[Z_n \mid Z_n > 0] = \frac{E[Z_n]}{P(Z_n > 0)} \sim \frac{\sigma^2}{2} n.
$$

Thus, the critical process can become very large, even though its extinction is certain. In theorem 1.5, we will see that $2Z_n/(\sigma^2 n)$, given non-extinction, converges in distribution to the unit Exponential distribution.

THEOREM 1.5.– *Let* $\{Z_n\}_{n\geq 0}$ *be a critical GW process with* $\sigma^2 < \infty$. *Then*

$$\lim_{n\to\infty} P\left(\frac{2Z_n}{\sigma^2 n} \leq x \,\middle|\, Z_n > 0\right) = P(Z \leq x), \qquad x \geq 0, \qquad [1.13]$$

where Z *has an Exponential distribution with mean 1.*

COMMENT.– This is another Yaglom-type limit theorem (see [YAG 47]).

Proof. Since the Laplace transform of the unit exponential distribution equals $(1+\lambda)^{-1}$, $\lambda > 0$, according to the continuity theorem for Laplace transforms (see theorem A2.4 in appendix 2), it is sufficient to prove

$$\lim_{n\to\infty} E\left[\exp\left\{-\lambda\frac{2Z_n}{\sigma^2 n}\right\} \,\middle|\, Z_n > 0\right] = \frac{1}{1+\lambda} \qquad \lambda > 0. \qquad [1.14]$$

In [1.9], substituting s for $\exp\{-2\lambda/\sigma^2 n\}$, we have

$$E\left[\exp\left\{-\lambda\frac{2Z_n}{\sigma^2 n}\right\} \,\middle|\, Z_n > 0\right] = 1 - \frac{1}{1-f_n(0)}\left[1 - f_n\left(\exp\left\{-\frac{2\lambda}{\sigma^2 n}\right\}\right)\right]. \qquad [1.15]$$

We start with finding the limit of the numerator in the right-hand side of the equation. Define

$$r := r(n) = \max\left\{k : \; f_k(0) \leq \exp\left\{-\frac{2\lambda}{\sigma^2 n}\right\}\right\}.$$

Therefore,

$$f_r(0) \leq \exp\left\{-\frac{2\lambda}{\sigma^2 n}\right\} \leq f_{r+1}(0) \qquad [1.16]$$

and thus,

$$1 - f_{r+1}(0) \leq 1 - \exp\left\{-\frac{2\lambda}{\sigma^2 n}\right\} \leq 1 - f_r(0).$$

This is equivalent to

$$1 - f_{r+1}(0) \leq \frac{2\lambda}{\sigma^2 n}(1 + \varepsilon_n) \leq 1 - f_r(0),$$

where $\varepsilon_n \to 0$ as $n \to \infty$. On the other hand, from [1.8] as $r \to \infty$

$$1 - f_r(0) = P(Z_r > 0) = \frac{2\lambda}{\sigma^2 r}(1 + o(1)).$$

Thus, $r \sim [n/\lambda]$ as $n \to \infty$, with $[x]$ denoting the integer part of x. Recalling [1.16], we have

$$1 - f_n(f_{r+1}(0)) \leq 1 - f_n\left(\exp\left\{-\frac{2\lambda}{\sigma^2 n}\right\}\right) \leq 1 - f_n(f_r(0)).$$

Since $r \sim [n/\lambda]$ as $n \to \infty$, applying again [1.8], we obtain as $n \to \infty$

$$\begin{aligned} 1 - f_n\left(\exp\left\{-\frac{2\lambda}{\sigma^2 n}\right\}\right) &\sim 1 - f_{n+r}(0) \\ &\sim \frac{2}{\sigma^2(n+r)} \\ &\sim \frac{2\lambda}{\sigma^2 n(1+\lambda)}. \end{aligned} \qquad [1.17]$$

It follows from [1.8] and [1.17] that as $n \to \infty$

$$1 - \frac{1 - f_n\left(\exp\{-\frac{2\lambda}{\sigma^2 n}\}\right)}{1 - f_n(0)} \to 1 - \frac{\lambda}{1+\lambda} = \frac{1}{1+\lambda},$$

which, taking into account [1.15], implies [1.14]. ∎

According to [1.13], if $\sigma^2 < \infty$ then Z_n, given non-extinction, converges at the rate n to an exponential distribution. What happens if the condition $\sigma^2 < \infty$ is replaced with the assumption that the offspring distribution is in the domain of attraction of a stable law with parameter α such that $0 < \alpha \leq 1$? In this case, it can be shown that, given non-extinction, Z_n grows at rate $n^{1/\alpha}$. More precisely, as it was proven in [SLA 68], if

$$f(s) = s + (1-s)^{1+\alpha} L(1/(1-s)), \qquad \alpha \in (0,1], \tag{1.18}$$

where $L(t)$ is a slowly varying function (s.v.f.) as $t \to \infty$ (see Appendix 2, definition A2.1), then for any $\lambda > 0$

$$\lim_{n\to\infty} E\left[\exp\left\{-\lambda \frac{Z_n}{n^{1/\alpha} L_1(n)}\right\} \;\Big|\; Z_n > 0\right] = 1 - \lambda(1+\lambda^{\alpha})^{-1/\alpha},$$

where $L_1(n)$ is a certain s.v.f. at infinity.

Finally, consider **the supercritical** $\{Z_n\}_{n\geq 0}$. The process survives with positive probability and if the population does not die out, it diverges to infinity. Also, recall that, as $n \to \infty$

$$E[Z_n] \sim m^n \qquad \text{and} \qquad CV[Z_n] \to \frac{\sigma}{\sqrt{m(m-1)}} < \infty.$$

These facts suggest that, after scaling with its expected value m^n, the population size Z_n will converge in some sense. We will prove that there exists a random variable W, such that $\lim_{n\to\infty} Z_n/m^n = W$ almost surely (a.s.). Thus, the random variable Z_n grows like $m^n W$. This is the stochastic analog of the so-called Malthusian law of geometric population growth. Note that, unlike the law of large numbers where (after scaling by the expected value) the limit is a constant, for the supercritical GW process, the limit W is a random variable.

THEOREM 1.6.– *Let $\{Z_n\}_{n\geq 0}$ be a supercritical GW process with $\sigma^2 < \infty$. Then there exists a random variable W, such that, as $n \to \infty$,*

$$W_n := \frac{Z_n}{m^n} \to W \ a.s. \tag{1.19}$$

Furthermore,

1) $\lim_{n\to\infty} E\left[(W_n - W)^2\right] = 0$, $E[W] = 1$, *and* $Var[W] = \sigma^2/(m^2 - m)$;

2) $P(W = 0) = q = P(extinction)$;

3) $\phi(\lambda) := E[\exp\{-\lambda W\}]$ *is the unique strictly monotone increasing and concave solution of the Poincaré equation for* $\lambda > 0$

$$\phi(m\lambda) = f(\phi(\lambda)), \qquad \phi(0) = 1. \tag{1.20}$$

COMMENT.– A stronger result (Kesten–Stigum theorem) states that [1.19] holds if and only if $E[X_{0,1}\log(1 + X_{0,1})] < \infty$. The latter condition is known as "$x \log x$ condition". A further refinement is that there exists a sequence of positive constants $\{C_n\}_{n\geq 0}$ (called Seneta constants), such that $C_{n+1}/C_n \to m$ and $Z_n/C_n \to W$ a.s., $P(W = 0) = q$, and W has a continuous distribution on the set of positive real numbers.

Proof. The main argument in the proof is based on the fact that $\{W_n, \mathcal{F}_n\}_{n\geq 0}$ with $\mathcal{F}_n = \sigma(Z_0, Z_1, \ldots, Z_n)$ is a martingale (see Appendix 1). Indeed,

$$\begin{aligned} E[W_n \mid \mathcal{F}_{n-1}] &= E\left[\frac{Z_n}{m^n} \,\middle|\, Z_{n-1}\right] \\ &= \frac{1}{m^n} E[Z_n \mid Z_{n-1}] \\ &= \frac{1}{m^n} E\left[\sum_{k=1}^{Z_{n-1}} X_{n-1,k} \,\middle|\, Z_{n-1}\right] \\ &= \frac{1}{m^n} Z_{n-1} m = W_{n-1}. \end{aligned}$$

In addition, $E[W_n] = 1$. Therefore, $\{W_n, \mathcal{F}_n\}_{n\geq 0}$ is a non-negative martingale. It follows from the martingale convergence theorem (see Appendix 1), that there exists a random variable W such that $\lim_{n\to\infty} W_n = W$ a.s. Using theorem 1.2, we obtain

$$E\left[W_n^2\right] = \frac{E\left[Z_n^2\right]}{m^{2n}} = \frac{\sigma^2(1 - m^{-n})}{m^2 - m} + 1.$$

Therefore,

$$\sup_n E\left[W_n^2\right] = \lim_{n\to\infty} E\left[W_n^2\right] = \frac{\sigma^2}{m^2 - m} + 1 < \infty.$$

Now, claim 1) follows by the properties of martingales and Doob's theorem (see Appendix 1).

To prove claim 2), first note that since $E[W] = 1$, we have $P(W = 0) < 1$. Using the branching property, we obtain

$$\begin{aligned} P(W = 0) &= \sum_{k=0}^{\infty} P(W = 0 \mid Z_1 = k)P(Z_1 = k) \\ &= \sum_{k=0}^{\infty} (P(W = 0 \mid Z_1 = 1))^k P(Z_1 = k) \\ &= f(P(W = 0)). \end{aligned}$$

Hence, $P(W = 0)$ is the smallest root of the equation $f(s) = s$, $s \in [0, 1]$, i.e. it is equal to the probability of extinction q.

To prove claim 3), observe that the fundamental equation [1.3], in terms of the Laplace transform $\varphi_n(u) := E[\exp\{-uZ_n\}]$, can be written as

$$\varphi_{n+1}(u) = f(\varphi_n(u)). \qquad [1.21]$$

On the other hand, if $\phi_n(u) := E[\exp\{-uW_n\}]$, then

$$\varphi_n(u) = E[\exp\{-uW_n m^n\}] = \phi_n(um^n).$$

Substituting this in [1.21] yields $\phi_{n+1}(um^{n+1}) = f(\phi_n(um^n))$. After change in variables $\lambda = um^n$, we have

$$\phi_{n+1}(m\lambda) = f\left(\phi_n(\lambda)\right).$$

Taking limit as $n \to \infty$, we obtain the Poincaré equation [1.20]. For a proof of the uniqueness of the solution, see [SEN 69]. ∎

1.2. Processes with unrestricted immigration

In this section, we will discuss GW processes, modified to allow immigration into the population. This process is defined by two distributions, an offspring distribution and an immigration distribution. More specifically, at time n a random number of new individuals ("immigrants" or "invaders") join the residents of the population. Once an individual enters the population, it reproduces according to the law of the underlying GW process. In general, the immigration may depend on the time n and/or the population size at time n, but must be independent of the reproduction of the individuals already in the population. We will focus our attention to some selected classes of branching processes with immigration. We would not be interested so much in presenting and proving the most general results available, but rather in a proof of a special case providing a background for the more general models to be considered in the following chapters.

As a rule, the proofs are based on a detailed asymptotic analysis of functional equations for the p.g.f.s of the processes. In addition, probabilistic constructions such as the one in the proof of theorem 1.10 are used.

Let us begin with GW processes subject to **unrestricted immigration** at every generation (time) n. The number of immigrants is independent of both n and population size at time n. More precisely, consider the offspring random variables $\{X_{n,i} : n = 0, 1, \ldots; i = 1, 2, \ldots\}$ with common distribution $\{p_k\}_{k\geq 0}$ defined earlier. Let on the same probability space be defined a set $\{I_n : n = 0, 1, \ldots\}$ of non-negative integer-valued i.i.d. random variables with common distribution $\{g_k\}_{k\geq 0}$ and independent from the offspring random variables. A GW process with immigration (GWI process) is defined as follows.

DEFINITION 1.4.– *The GWI process is a discrete time homogeneous Markov chain $\{Y_n\}_{n\geq 0}$ defined inductively by $Y_0 = 0$ and for $n = 0, 1, \ldots$*

$$Y_{n+1} = \begin{cases} \sum_{i=1}^{Y_n} X_{n,i} + I_{n+1} & \textit{if} \quad Y_n > 0 \\ I_{n+1} & \textit{if} \quad Y_n = 0. \end{cases} \qquad [1.22]$$

The random variable I_{n+1} represents the number of immigrants joining the population at the $(n + 1)$st generation (see Figure 1.3). If $g_0 = 1$, we recover the ordinary GW process.

Let $Z_n^{(i)}(k)$ be the size of a family, n generations after it was initiated by the ith immigrant in the group of immigrants I_k that arrived at time k. Then an alternative definition of a GWI process is: $Y_0 = 0$ and for $n = 0, 1, \ldots$

$$Y_{n+1} = \sum_{i=1}^{I_0} Z_n^{(i)}(0) + \sum_{i=1}^{I_1} Z_{n-1}^{(i)}(1) + \ldots + \sum_{i=1}^{I_n} Z_0^{(i)}(n). \qquad [1.23]$$

The decomposition [1.23] is a convenient tool in constructing continuous-time versions and other generalizations of GWI process by letting $Z_n^{(i)}(k)$ be a Markov branching process, a Bellman–Harris process, etc. (e.g. [ATH 72]).

For the p.g.f. $H_n(s)$ of Y_n we have $H_0(s) = 1$ and for $n = 1, 2, \ldots$

$$\begin{aligned} H_{n+1}(s) &= E\left[s^{Y_{n+1}} \mid Y_0 = 0\right] \qquad [1.24] \\ &= E\left[s^{\sum_{i=1}^{Y_n} X_{n,i} + I_{n+1}} \mid Y_0 = 0\right] \\ &= g(s) H_n(f(s)) \\ &= \prod_{k=0}^{n+1} g(f_k(s)), \end{aligned}$$

where $f(s)$ and $g(s)$ are the offspring and immigration p.g.f.s, respectively.

DEFINITION 1.5.– *A GWI process is said to be subcritical, critical, or supercritical if the offspring mean m satisfies $m < 1$, $m = 1$ or $m > 1$, respectively.*

That is, the process inherits the criticality of the resident population. To avoid trivialities, we will assume $p_0 > 0$, $p_0 + p_1 < 1$ and $g_0 < 1$. Similarly to theorem 1.2, differentiating the fundamental equation [1.24], one can calculate $E[Y_n]$ and, denoting $a := E[I_n] < \infty$, obtain as $n \to \infty$

$$E[Y_n] \sim \begin{cases} \dfrac{a}{m-1} m^n & \text{if} \quad m > 1 \\ an & \text{if} \quad m = 1 \\ \dfrac{a}{1-m} & \text{if} \quad m < 1. \end{cases} \qquad [1.25]$$

Therefore, if $m > 1$, then the rate of increase in the population size does not depend much on the immigration. If $m < 1$, the expected number of individuals remains bounded and if $m = 1$, it is proportional to n.

The following classification of the Markov chain $\{Y_n\}_{n\geq 0}$ holds. Assuming $a < \infty$ for $m \geq 1$ and denoting $\gamma := 2a/\sigma^2$, we have $\{Y_n\}_{n\geq 0}$ is

$$\begin{cases} \text{transient} & \text{if} \quad m > 1 \\ \text{transient} & \text{if} \quad m = 1,\ \sigma^2 < \infty, \quad \gamma > 1 \\ \text{null–recurrent} & \text{if} \quad m = 1,\ \sigma^2 < \infty, \quad \gamma = 1 \text{ and [1.27]} \\ \text{null–recurrent} & \text{if} \quad m = 1,\ \sigma^2 < \infty, \quad 0 \leq \gamma < 1 \\ \text{pos. recurrent} & \text{if} \quad m = 1,\ \sigma^2 = \infty \\ \text{pos. recurrent} & \text{if} \quad m < 1,\ E[\log(1 + I_1)] < \infty, \end{cases} \qquad [1.26]$$

where

$$E\left[X_{0,1}^2 \log(1 + X_{0,1})\right] < \infty \quad \text{and} \quad E[I_1 \log(1 + I_1)] < \infty. \qquad [1.27]$$

In the critical case, note the central role played by the parameter γ, which equals the average immigration divided by half of the offspring variance.

In contrast to the ordinary GW process, the state zero is a reflecting barrier for a GWI process. Therefore, instead of the extinction time T, a random variable of interest here is the duration τ of a **life-period**, i.e. the time between two consecutive visits to zero.

DEFINITION 1.6.– *The random variable τ is said to be (duration of) a life-period starting at n_0 if $Y_{n_0-1} = 0$, $Y_n > 0$ when $n_0 \leq n < n_0 + \tau$, and $Y_{n_0+\tau} = 0$.*

In the **critical case** with $\sigma^2 < \infty$ and $a < \infty$, it can be proven under some additional moment conditions (see [ZUB 72]) that as $n \to \infty$

$$P(\tau > n) \sim \begin{cases} c_1 & \text{for} \quad \gamma > 1 \\ c_2(\log n)^{-1} & \text{for} \quad \gamma = 1 \\ c_3 n^{-(1-\gamma)} & \text{for} \quad 0 \leq \gamma < 1, \end{cases} \qquad [1.28]$$

where $c_i := c_i(\gamma)$ for $1 \leq i \leq 3$ are positive constants. In the **subcritical case**, it can be proven (see [HAC 05, p. 181]) that if $a < \infty$

$$E[\tau] = \frac{(1 - r)g(0)}{r(1 - g(0))},$$

where $r := \lim_{n\to\infty} r_n$ and $r_n := \prod_{k=0}^{n} g(f_k(0))$. The absolute error in the approximation of r by r_n satisfies $0 \leq r_n - r \leq am^{n+1}/(1-m)$.

1.2.1. *Limit theorems*

It can be expected that the immigration will compensate the tendency to extinction when $m \leq 1$ and Y_n would converge in distribution. Theorem 1.7 shows that this is indeed true in the **subcritical case**.

THEOREM 1.7.– *Let $\{Y_n\}_{n\geq 0}$ be a subcritical GWI process. If $\{Y_n\}_{n\geq 0}$ is an irreducible and aperiodic Markov chain, then*

$$E[\log(1+I_1)] < \infty$$

is a necessary and sufficient condition for the existence of a limiting–stationary distribution $\{\pi_k\}_{k\geq 0}$ with p.g.f.

$$\sum_{k=0}^{\infty} \pi_k s^k = \prod_{k=0}^{\infty} g(f_k(s)), \qquad 0 \leq s \leq 1. \qquad [1.29]$$

COMMENT.– The significance of this result is that the exponential decline of the population size is balanced by the immigration if $E[\log(1+I_1)] < \infty$. We denote the one-step transition probability from state i to state j by p_{ij}, i.e. $p_{ij} := P(Y_{n+1} = j \mid Y_n = i)$. According to the Markov chain theory, if $E[I_1] = \infty$, then $\sum_{j=0}^{\infty} jp_{ij} = \infty$ for every i ; however, the chain is still positive recurrent provided $E[\log(1+I_1)] < \infty$.

The proof of theorem 1.7 is provided in [HEA 66]. We will show here that [1.29] holds (the sufficiency part) under the stronger assumption $a = E[I_1] < \infty$. In view of [1.24], it is sufficient to show that

$$\sum_{k=0}^{\infty} \pi_k s^k = \Pi(s) := \lim_{n\to\infty} H_n(s) = \prod_{k=0}^{\infty} g(f_k(s)) \qquad [1.30]$$

exists. Since $a = g'(1) < \infty$, we have

$$1 - g(f_k(s)) \leq g'(1)(1 - f_k(s)) \leq am^k(1-s).$$

Therefore, $\sum_{k=0}^{\infty}(1-g(f_k(s))) < \infty$, which implies $0 < \prod_{k=0}^{\infty} g(f_k(s)) \leq 1$ for all $s \in [0,1]$. Hence, $\Pi(1) = 1$ and [1.30] holds. According to [1.25], the limiting distribution has mean $a/(1-m)$.

A more general problem is the existence and uniqueness of an invariant measure for the Markov chain $\{Y_n\}_{n\geq 0}$, that is a non-negative sequence $\{\mu_i\}_{i\geq 0}$ ($\mu_i > 0$ for some i) such that for $j = 0, 1, \ldots$

$$\mu_j = \sum_{i=0}^{\infty} \mu_i p_{ij}. \quad [1.31]$$

It can be proven (see [SEN 69]) that if $m \leq 1$ and $U(s) := \sum_{k=0}^{\infty} \mu_k s^k$, then $U(s)$ is the unique (up to a constant multiplier) solution of the generalized Schröder functional equation (to be compared with [1.11])

$$U(x) = g(x)U(f(x)), \qquad 0 \leq x < 1. \quad [1.32]$$

In **the supercritical case**, the immigration does not have much effect on the long-range evolution of the population. If $E[\log(1+I_1)] < \infty$, then there is a strictly increasing sequence of constants $\{c_n\}_{n\geq 0}$ and $c_{n+1}/c_n \rightarrow m$ for which $\{Y_n/c_n\}_{n\geq 0}$ converges almost surely to a non-degenerate positive random variable $\tilde{V}$ with continuous cumulative distribution function (c.d.f.). Let q be the least non-negative solution of the equation $f(s) = s$ and $\delta = -\log g(q)/\log m$. The following is known (see [PAK 74]) about the limiting variable $\tilde{V}$.

THEOREM 1.8.– *Let $\{Y_n\}_{n\geq 0}$ be a supercritical GWI process. The following claims are then true.*

1) The limit $\tilde{V}$ has an absolutely continuous c.d.f.

2) If $\delta > 1$, then $\tilde{V}$ has a bonded continuous density function.

3) If $E[I_1] < \infty$ and $E[X_{0,1}\log(1+X_{0,1})] < \infty$, then $\tilde{V}$ has a density function that is continuous on the set of positive real numbers.

We proceed to **the critical GWI process**. The immigration provides an appealing alternative to conditioning on non-extinction as theorem 1.9 shows.

THEOREM 1.9.– *Let* $\{Y_n\}_{n\geq 0}$ *be a critical GWI process with finite offspring variance* $\sigma^2 < \infty$. *If* $a < \infty$, *then*

$$\lim_{n\to\infty} P\left(\frac{2Y_n}{\sigma^2 n} \leq y\right) = P(Y \leq y), \qquad [1.33]$$

where the limit Y *has Gamma distribution with parameters* γ *and 1.*

COMMENT.– (i) If $\gamma = 2a/\sigma^2 = 1$, then the limit in [1.33] is unit exponential as in theorem 1.5. Hence, the conditioning on non-extinction in the ordinary GW process has the same effect as allowing immigration at an average rate of $a = \sigma^2/2$. (ii) Further results assuming [1.18] can be found in [PAK 79].

Proof. Taking Laplace transform in each side of [1.33], we see that it is sufficient to prove

$$\lim_{n\to\infty} E\left[\exp\left\{-\lambda\frac{2Y_n}{\sigma^2 n}\right\}\right] = \frac{1}{(1+\lambda)^{\gamma}}, \qquad \lambda > 0. \qquad [1.34]$$

As in the proof of theorem 1.5, there exists $r := r(n)$, such that

$$f_r(0) \leq \exp\left\{-\frac{2\lambda}{\sigma^2 n}\right\} \leq f_{r+1}(0) \qquad [1.35]$$

and $r \sim [n/\lambda]$ as $n \to \infty$. Combining [1.35] and

$$E\left[\exp\left\{-\lambda\frac{2Y_n}{\sigma^2 n}\right\}\right] = \prod_{k=0}^{n} g\left(f_k\left(\exp\left\{-\frac{2\lambda}{\sigma^2 n}\right\}\right)\right),$$

we have

$$\begin{aligned}\prod_{k=r}^{n+r} g(f_k(0)) &= \prod_{j=0}^{n} g(f_j(f_r(0))) \qquad [1.36]\\ &\leq \prod_{j=0}^{n} g\left(f_j\left(\exp\left\{-\frac{2\lambda}{\sigma^2 n}\right\}\right)\right)\\ &\leq \prod_{k=r+1}^{n+r+1} g(f_k(0)).\end{aligned}$$

Since the ratio of the upper and lower bound $g(f_{n+r+1}(0))/g(f_r(0)) \to 1$ as $r \to \infty$, they both have the same limit. We will find the limit of the lower bound. Taking into account that as $s \uparrow 1$

$$g(s) = 1 - a(1-s) + o(1-s) = \exp\{-a(1-s) + o(1-s)\},$$

we obtain as $n \to \infty$

$$\begin{aligned}
\prod_{k=r}^{n+r} g(f_k(0)) &= \exp\left\{-a\sum_{k=r}^{n+r}(1-f_k(0))(1+o(1))\right\} \\
&= \exp\left\{-\frac{2a}{\sigma^2}\sum_{k=r}^{n+r}\frac{1}{k}(1+o(1))\right\} \\
&= \exp\left\{-\frac{2a}{\sigma^2}\log\frac{n+r}{r}(1+o(1))\right\} \\
&= \exp\left\{-\gamma\log\frac{n+n/\lambda}{n/\lambda}(1+o(1))\right\} \\
&= \exp\{-\gamma\log(1+\lambda)(1+o(1))\} \\
&\to \frac{1}{(1+\lambda)^\gamma}.
\end{aligned}$$

Since the upper bound in [1.36] approaches the same limit, [1.34] follows by [1.36] and the Sandwich theorem. ∎

Theorem 1.9 can be interpreted as follows. At each time n, a immigrants, on average, join the population. The probability of one immigrant being "successful" n generation later is $P(Z_n > 0) \sim 2/(\sigma^2 n)$. The entire immigration group at time n will be successful with probability $2a/(\sigma^2 n)$. Therefore, on average, $\gamma = 2a/\sigma^2$ number of processes will survive in the long run. Since, from theorem 1.5, the size of each process, divided by $\sigma^2 n/2$ approaches an Exponential variable with mean one, we have that $2Y_n/(\sigma^2 n)$ approaches a Gamma variable with mean γ.

1.2.2. *Critical process with decreasing to zero immigration*

Let us consider a critical GWI process with time–depending and decreasing to zero immigration. That is, the immigration rate at each generation time is decreasing to zero. More precisely, let $Y_0 = 0$ and for $n = 0, 1, \dots$

$$Y_{n+1} = \sum_{i=1}^{Y_n} X_{n,i} + I_{n+1},$$

(with $\sum_1^0 = 0$), where the random variables $\{I_n\}_{n\geq 1}$ are not identically distributed and $a_n := E[I_n]$ converges to zero as $n \to \infty$. Define $g_n(s) := E[s^{I_n}]$ and $c_n := g_n''(1)$. Suppose $\sup_n a_n < \infty$ and $\sup_n c_n < \infty$. It can be expected that the contribution of the immigration to the limiting behavior of $\{Y_n\}_{n\geq 0}$ depends on how fast the immigration rate a_n converges to zero.

Similarly to [1.24], for the p.g.f. of the process we obtain the equation

$$H_n(s) := E\left[s^{Y_n}\right] = \prod_{k=0}^{n} g_k(f_{n-k}(s)), \qquad H_0(s) = 1.$$

Let us weaken the condition $\sigma^2 < \infty$ and assume instead that (see also [1.18])

$$f(s) = s + (1-s)^{1+\alpha} L_f(1/(1-s)), \qquad \alpha \in (0, 1], \tag{1.37}$$

where $L_f(t)$ is an s.v.f. as $t \to \infty$. Also for the immigration mean assume

$$a_n = \begin{cases} n^{-\rho} L_a(n), \quad L_a(n) \text{ - s.v.f. at } \infty & \text{for } \rho > 0 \\ L_a(n) = o\left(\left(\sum_{k=0}^{n}(1 - f_k(0))\right)^{-1}\right) & \text{for } \rho = 0. \end{cases} \tag{1.38}$$

Define

$$Q_a(n) := a_n \sum_{k=0}^{n} (1 - f_k(0)),$$

$$Q_f(n) := (1 - f_n(0)) \sum_{k=0}^{n} a_k, \quad \theta_n := \frac{Q_a(n)}{Q_f(n)}.$$

THEOREM 1.10.– *Let* $\{Y_n\}_{n\geq 0}$ *be a GWI process with time-dependent immigration and offspring p.g.f.* $f(s)$ *satisfying* [1.37]. *Assume* [1.38] *and* $c_n = o(Q_a(n))$ *as* $n \to \infty$. *Then*

1) If $\liminf_{n\to\infty} \theta_n > 0$, *then for* $0 \leq x \leq 1$

$$\lim_{n\to\infty} P\left(\frac{L(Y_n)}{L\left((1-f_n(0))^{-1}\right)} \leq x \,\middle|\, Y_n > 0\right) = x \lim_{n\to\infty} \frac{\theta_n}{\theta_n + 1}. \qquad [1.39]$$

2) If $\limsup_{n\to\infty} \theta_n < \infty$ *and* $c_n = O(1)$, *then for* $x \geq 0$

$$\lim_{n\to\infty} P\left(\frac{Y_n}{(1-f_n(0))^{-1}} \leq x \,\middle|\, Y_n > 0\right) = \lim_{n\to\infty} \frac{\theta_n}{\theta_n+1} + \lim_{n\to\infty} \frac{1-e^{-x}}{\theta_n+1}, \qquad [1.40]$$

where $L(x)$ *is a certain slowly varying at infinity function (see [RAH 86], Theorem 2.3).*

Proof. (See [RAH 86] for a more detailed proof.) The proof can be divided into three steps as follows: (i) the GWI process $\{Y_n\}_{n\geq 0}$ is split into three independent ordinary GW processes called "partial processes"; (ii) using analytic methods, limit theorems are established for the partial processes; (iii) using probabilistic arguments the asymptotic results for the partial processes are transferred to the GWI process $\{Y_n\}_{n\geq 0}$.

Let $Z_n(k)$ be the number of descendants at time n of the group of immigrants which arrived at k (see Figure 1.3). Define positive and integer-valued functions $r_i(n)$ for $i = 1, 2$ such that $r_i(n) = o(n) \to \infty$ as $n \to \infty$ and

$$\sum_{k=0}^{r_1(n)} a_k \sim \sum_{k=0}^{n} a_k, \quad \rho \geq 1 \quad \text{and} \quad \sum_{k=0}^{r_2(n)} (1 - f_k(0)) \sim \sum_{k=0}^{n} (1 - f_k(0)).$$

By the assumptions of the theorem, functions with these properties exist and the following decomposition holds (see also [1.23]):

$$\begin{aligned} Y_n &= \sum_{k=0}^{n} Z_n(k) \qquad [1.41] \\ &= \sum_{k=0}^{r_1(n)} Z_n(k) + \sum_{k=r_1(n)+1}^{n-r_2(n)} Z_n(k) + \sum_{k=n-r_2(n)+1}^{n} Z_n(k) \end{aligned}$$

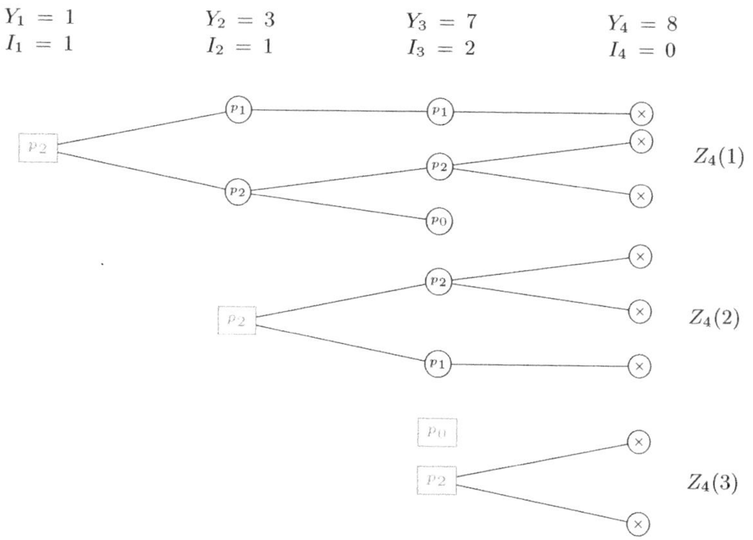

Figure 1.3. *Decomposition of GWI process*

$$=: Z_{n0} + Z_{n1} + Z_{n2}, \qquad \text{say}.$$

Define the following mutually exclusive events for $i,\ j,\ k = 0, 1, 2;\ i \neq j \neq k$

$$\begin{aligned} E_i &:= \{Z_{ni} > 0,\ Z_{nj} = 0,\ Z_{nk} = 0\}, \\ E_{ij} &:= \{Z_{ni} > 0,\ Z_{nj} > 0,\ Z_{nk} = 0\}, \\ E_{012} &:= \{Z_{n0} > 0,\ Z_{n1} > 0,\ Z_{n2} > 0\}. \end{aligned}$$

Clearly,

$$\{Y_n > 0\} = \left(\bigcup_{i=0}^{2} E_i\right) \bigcup \left(\bigcup_{i,j=0;\ i\neq j}^{2} E_{ij}\right) \bigcup E_{012}.$$

Therefore, for any event A, the total probability rule implies

$$P(A \mid Y_n > 0)P(Y_n > 0) = \sum_{i=0}^{2} P(A \mid E_i)P(E_i)$$

$$+ \sum_{i,j=0\ i\neq j}^{2} P(A \mid E_{ij})P(E_{ij}) + P(A \mid E_{012})P(E_{012}).$$

Furthermore, by the definition of the partial processes $\{Z_{n,i}\}_{n\geq 0}$ for $i = 0, 1, 2$, we have for $i,\ j,\ k = 0, 1, 2$ and $i \neq j \neq k$

$$\begin{aligned} P(E_i) &= P(Z_{ni} > 0)\ P(Z_{nj} = 0)\ P(Z_{nk} = 0) \\ P(E_{ij}) &= P(Z_{ni} > 0)\ P(Z_{nj} > 0)\ P(Z_{nk} = 0) \\ P(E_{012}) &= P(Z_{n0} > 0)\ P(Z_{n1} > 0)\ P(Z_{n2} > 0). \end{aligned} \quad [1.42]$$

Under the assumptions of the theorem, it can be proven that as $n \to \infty$

$$P(E_0) \sim Q_f(n),\ \ \rho \geq 1 \quad \text{and} \quad P(E_2) \sim Q_a(n).$$

and the rest of the events in [1.42] have probabilities $o(P(E_0) + P(E_2))$ as $n \to \infty$. Also, under the assumptions of the theorem, as $n \to \infty$

$$P(Y_n > 0) \sim Q_f(n) + Q_a(n).$$

Therefore, as $n \to \infty$

$$\begin{aligned} P(A \mid Y_n > 0) &= \left[\frac{P(A \mid E_0)P(E_0) + P(A \mid E_2)P(E_2)}{P(Y_n > 0)}\right](1 + o(1)) \\ &\sim \frac{P(A \mid E_0)Q_f(n) + P(A \mid E_2)Q_a(n)}{Q_f(n) + Q_a(n)} \end{aligned} \quad [1.43]$$

To prove claim 1) observe that for $i = 0, 1, 2$ and $0 \leq x \leq 1$

$$\begin{aligned} P\left(\frac{L(Y_n)}{L\left((1 - f_n(0))^{-1}\right)} \leq x \,\middle|\, E_i\right) &= P\left(\frac{L(Z_{n0} + Z_{n1} + Z_{n2})}{L\left((1 - f_n(0))^{-1}\right)} \leq x \,\middle|\, E_i\right) \\ &= P\left(\frac{L(Z_{ni})}{L\left((1 - f_n(0))^{-1}\right)} \leq x \,\middle|\, Z_{ni} > 0\right). \end{aligned} \quad [1.44]$$

On the other hand, studying the partial processes in detail, one can obtain

$$\lim_{n\to\infty} P\left(\frac{L(Z_{ni})}{L\left((1 - f_n(0))^{-1}\right)} \leq x \,\middle|\, Z_{ni} > 0\right) = \begin{cases} 0 \text{ for } & i = 0 \\ x \text{ for } & i = 2. \end{cases}$$

Substituting $A = \{L(Y_n)/L((1-f_n(0))^{-1}) \leq x\}$ in [1.43] we have as $n \to \infty$

$$P\left(\frac{L(Y_n)}{L((1-f_n(0))^{-1})} \leq x \,\middle|\, Y_n > 0\right) \sim \frac{Q_a(n)x}{Q_f(n)+Q_a(n)}$$

$$\sim \frac{\theta_n}{\theta_n+1}\,x,$$

which yields the claim 1).

We turn to the proof of claim 2). Similarly to [1.44] we have for $i = 0, 1, 2$

$$P\left(\frac{Y_n}{(1-f_n(0))^{-1}} \leq x \,\middle|\, E_i\right) = P\left(\frac{Z_{ni}}{(1-f_n(0))^{-1}} \leq x \,\middle|\, Z_{ni} > 0\right).$$

For the partial processes, it can be proven that for $x \geq 0$

$$\lim_{n\to\infty} P\left(\frac{Z_{ni}}{(1-f_n(0))^{-1}} \leq x \,\middle|\, Z_{ni} > 0\right) = \begin{cases} 1-e^{-x} & \text{for } i=0 \\ 1 & \text{for } i=2. \end{cases}$$

Substituting $A = \{Y_n/(1-f_n(0))^{-1} \leq x\}$ in [1.43] we have as $n \to \infty$

$$P\left(\frac{Y_n}{(1-f_n(0))^{-1}} \leq x \,\middle|\, Y_n > 0\right) \sim \frac{Q_a(n)+Q_f(n)(1-e^{-x})}{Q_f(n)+Q_a(n)}$$

$$\sim \frac{\theta_n}{\theta_n+1} + \frac{1-e^{-x}}{\theta_n+1},$$

which implies the claim in 2). ∎

COMMENT.– According to [1.39] and [1.40], Y_n, given $Y_n > 0$, has different limiting distributions under different normalizations. The decomposition [1.41] of $\{Y_n\}_{n\geq 0}$ into partial processes shows that $\{Y_n\}_{n\geq 0}$ survives in a long run due to the descendants of either early immigrants or most recent ones. The asymptotic behavior these two groups descendants yields [1.39] and [1.40].

1.3. Processes with immigration after empty generation only

Consider a GW process allowing immigration only when there are no individuals left in the population (process with "resurrection"). The following motivation is given in [PAK 71a]. An important feature of the GWI process from the previous section is that no matter how large the size of a given generation, immigration may still take place in the following generation. It is not difficult to conceive of the possibility where, if the size of a generation exceeds some pre assigned number, immigration is not allowed in the following generation. In the simplest such process, immigration is allowed in a generation if and only if the previous generation was empty.

DEFINITION 1.7.– *The GW process with immigration at zero only is a discrete time homogeneous Markov chain* $\{Y_n^0\}_{n\geq 0}$, *a.k.a. Foster–Pakes (FP) process, defined by:* $Y_0^0 = 0$ *and for* $n = 0, 1, \ldots$

$$Y_{n+1}^0 = \begin{cases} \sum_{i=1}^{Y_n^0} X_{n,i} & \text{if } \; Y_n^0 > 0 \\ \\ I_{n+1} & \text{if } \; Y_n^0 = 0. \end{cases} \qquad [1.45]$$

The "dual" process with immigration "stopped at zero", i.e. with immigration at all states but zero, is of independent interest, but will not be discussed here.

Recall that $f(s) = E\left[s^{X_{0,1}}\right]$, $g(s) = E\left[s^{I_1}\right]$. The p.g.f. of the process $H_n^0(s) := E\left[s^{Y_n^0}\right]$ satisfies the following equation

$$\begin{aligned} H_{n+1}^0(s) &= E\left[s^{\sum_{i=1}^{Y_n^0} X_{n,i} + I_{n+1}\mathcal{I}\{Y_n^0=0\}}\right] \qquad [1.46] \\ &= H_n^0(f(s)) - H_n^0(0) + H_n^0(0)g(s) \\ &= H_n^0(f(s)) - (1 - g(s))H_n^0(0) \\ &= 1 - \sum_{k=0}^{n} (1 - g(f_k(s)))H_{n-k}^0(0), \end{aligned}$$

where $\mathcal{I}A$ is the indicator of the set A. Differentiating [1.46] and substituting $s = 1$, we have

$$E\left[Y_{n+1}^0\right] = a \sum_{k=0}^{n} m^{n-k} H_k^0(0). \qquad [1.47]$$

DEFINITION 1.8.– *The process* $\{Y_n^0\}_{n\geq 0}$ *is said to be subcritical, critical or supercritical as the offspring mean* m *satisfies* $m < 1$, $m = 1$ *or* $m > 1$, *respectively.*

1.3.1. *Limit theorems*

In the **subcritical case**, a limiting–stationary distribution exists for the process $\{Y_n^0\}_{n\geq 0}$ as it is shown in theorem 1.11.

THEOREM 1.11.– *Let* $\{Y_n^0\}_{n\geq 0}$ *be a subcritical FP process. If* $\{Y_n^0\}_{n\geq 0}$ *is irreducible and aperiodic, then*

$$E[\log(1+I_1)] < \infty$$

is a necessary and sufficient condition for the existence of a limiting–stationary distribution $\{\pi_k\}_{k\geq 0}$ *with p.g.f.*

$$\sum_{k=0}^{\infty} \pi_k s^k = 1 - \pi_0 \sum_{k=0}^{\infty} (1 - g(f_k(s))),\ \pi_0 = \left(1 + \sum_{k=0}^{\infty} (1 - g(f_k(0)))\right)^{-1}. \tag{1.48}$$

COMMENT.– A sufficient condition for irreducibility and aperiodicity is for p_1 to be positive.

The proof of theorem 1.11 is provided in [PAK 71a]. We will prove the existence of a limiting–stationary distribution [1.48] (the sufficiency part) under the stronger assumption $E[I_1] < \infty$. Let $p_{ij}^{(n)}$ be the n-step transition probability from state i to state j, i.e. $p_{ij}^{(n)} := P(Y_{k+n}^0 = j \mid Y_k^0 = i)$. It is known (e.g. [FEL 66, p. 356]) that if a Markov chain is irreducible and aperiodic, then either

(i) for any pair of states (i,j), the probability $\lim_{n\to\infty} p_{ij}^{(n)} = 0$ and therefore there is no stationary distribution, or

(ii) all the states are ergodic, i.e. $\lim_{n\to\infty} p_{ij}^{(n)} = h_j > 0$. In this case, $\{h_j\}_{j\geq 0}$ is a unique stationary distribution.

Now, for the FP process $\{Y_n^0\}_{n\geq 0}$, consider the n step transition probability $p_{00}^{(n)} = H_n^0(0)$. Assuming there is no stationary distribution, we obtain (by dominated convergence theorem) the contradiction

$$0 = \lim_{n\to\infty} H_{n+1}^0(0) = 1 - \lim_{n\to\infty} \sum_{k=0}^{n} (1 - g(f_k(0))) H_{n-k}^0(0) = 1,$$

since for the series in the right-hand side of the equation we have

$$\sum_{k=0}^{\infty}(1 - g(f_k(0))) \leq g'(1)\sum_{k=0}^{\infty}(1 - f_k(0)) \leq g'(1)\sum_{k=0}^{\infty} m^k < \infty.$$

Thus, there exists a stationary distribution, which p.g.f. $\Pi^0(s)$, say, satisfies $\lim_{n\to\infty} H_n^0(s) = \Pi^0(s)$. Furthermore, [1.46] yields

$$\Pi^0(s) = \Pi^0(f(s)) - \pi_0(1 - g(s)) = 1 - \pi_0 \sum_{k=0}^{\infty}(1 - g(f_k(s))). \quad [1.49]$$

Therefore,

$$\pi_0 = 1 - \pi_0 \sum_{k=0}^{\infty}(1 - g(f_k(0))),$$

which, in view of [1.49], completes the proof. When $a < \infty$, the mean of the stationary distribution is $a\pi_0/(1-m)$.

It can be proven (see [PAK 71a]) that if $m \leq 1$ and $U(s) := \sum_{k=0}^{\infty} \mu_k s^k$, where $\{\mu_k\}_{k\geq 0}$ is an invariant measure for $\{Y_n^0\}_{n\geq 0}$ (see [1.31]), then $U(s)$ is the unique (up to a constant multiplier) solution of the functional equation (to be compared with [1.32])

$$U(x) = U(f(x)) - (1 - g(x)), \qquad 0 \leq x < 1; \quad \mu_0 = 1. \qquad [1.50]$$

In **the supercritical case**, the limiting behavior of the process is similar to that for a GW process. Namely, if $m > 1$ and $E[I_1] < \infty$, then there is a strictly increasing sequence of constants $\{c_n\}_{n\geq 0}$ and $c_{n+1}/c_n \to m$ for which $\{Y_n/c_n\}_{n\geq 0}$ converges almost surely to a non-degenerate positive random variable $\tilde{V}^0$ with continuous c.d.f. If, furthermore, $E[I_1 \log(1 + I_1)] < \infty$, then c_n can be replaced by m^n. Let q be the least non-negative solution of $f(s) = s$. If $f'(q) > 0$ set $\nu = -\log f'(q)/\log m$ and $\nu = \infty$ if $f'(q) = 0$. The limiting variable $\tilde{V}^0$ has the following properties.

THEOREM 1.12.– *Let $\{Y_n^0\}_{n\geq 0}$ be a supercritical FP process. Then the following claims are true.*

1) The limiting random variable $\tilde{V}^0$ has an absolutely continuous c.d.f.

2) If $\nu > 1$, then $\tilde{V}^0$ has a bounded continuous density function.

3) If $E[I_1] < \infty$ and $E[X_{0,1}\log(1 + X_{0,1})] < \infty$, then $\tilde{V}^0$ has a density function that is continuous on the set of positive real numbers.

Let us discuss in more detail the **critical case** with finite offspring variance $\sigma^2 < \infty$. Analyzing the sum in [1.47], one can obtain that as $n \to \infty$

$$E\left[Y_{n+1}^0\right] \sim \frac{\sigma^2}{2}\frac{n}{\log n} \quad \text{and} \quad E\left[(Y_{n+1}^0)^2\right] \sim \frac{\sigma^4}{4}\frac{n^2}{\log n}.$$

It can also be proven (see [FOS 71]) that as $n \to \infty$

$$H_n^0(0) = P\left(Y_n^0 = 0\right) \sim \frac{\sigma^2}{2a}\frac{1}{\log n}. \qquad [1.51]$$

THEOREM 1.13.– *Let $\{Y_n^0\}_{n\geq 0}$ be a critical FP process with $\sigma^2 < \infty$ and $a < \infty$. Then, for $0 < x < 1$,*

$$\lim_{n\to\infty} P\left(\frac{\log Y_n^0}{\log n} \leq x\right) = P\left(Y^0 \leq x\right), \qquad [1.52]$$

where the limit Y^0 has Uniform distribution over the interval $(0, 1)$.

Proof. We need the following Spitzer's comparison lemma (see [ATH 72, p. 22]). Under the assumptions of the theorem, for every $\varepsilon > 0$, there is a number $s_0 \in (0, 1)$, such that for any $k \geq 0$ and for all $s \in [s_0, 1)$,

$$\left[\frac{1}{1-s} + \frac{k}{2}(\sigma^2 + \varepsilon)\right]^{-1} \leq 1 - f_k(s) \leq \left[\frac{1}{1-s} + \frac{k}{2}(\sigma^2 - \varepsilon)\right]^{-1}. \quad [1.53]$$

Also, under the assumptions of the theorem, for the immigration p.g.f., we have that for every $\varepsilon \in (0, a)$ there exists $t_0 \in (0, 1)$ such that for all $s \in [t_0, 1)$

$$(a - \varepsilon)(1 - s) \leq 1 - g(s) \leq (a + \varepsilon)(1 - s). \qquad [1.54]$$

Combining [1.53] and [1.54], and recalling the equation [1.46], we obtain for fixed $s \in [t_0, 1)$, $x \in (0, 1)$ and n sufficiently large

$$H_{n+1}^0\left(e^{-s/n^x}\right) \leq 1 - \sum_{k=0}^{n}(a - \varepsilon)\left[\frac{2n^x}{s} + \frac{k}{2}(\sigma^2 + \varepsilon)\right]^{-1} H_{n-k}^0(0)$$

and

$$H_{n+1}^0\left(e^{-s/n^x}\right) \geq 1 - \sum_{k=0}^{n}(a-\varepsilon)\left[\frac{n^x}{s}+\frac{k}{2}(\sigma^2-\varepsilon)\right]^{-1} H_{n-k}^0(0),$$

where we have used that for u sufficiently small $u/2 \leq 1 - \exp\{-u\} \leq u$. Recalling the asymptotic result [1.51], we have

$$1 - \frac{\sigma^2}{2}\lim_{n\to\infty}\sum_{k=0}^{n-2}\left[\frac{n^x}{s}+\frac{k}{2}(\sigma^2-\varepsilon)\right]^{-1}[\log(n-k)]^{-1} \qquad [1.55]$$

$$\leq \liminf_{n\to\infty} H_{n+1}^0\left(e^{-s/n^x}\right) \leq \limsup_{n\to\infty} H_{n+1}^0\left(e^{-s/n^x}\right)$$

$$\leq 1 - \frac{\sigma^2}{2}\left(1-\frac{\varepsilon}{a}\right)\lim_{n\to\infty}\sum_{k=0}^{n-2}\left[\frac{2n^x}{s}+\frac{k}{2}(\sigma^2+\varepsilon)\right]^{-1}[\log(n-k)]^{-1}$$

Let us evaluate the limit as $n \to \infty$ of the following sum

$$\sum_{k=0}^{n-2}\frac{\sigma^2}{2}\frac{1}{n^x/s+k\sigma^2/2}\frac{1}{\log(n-k)} = \sum_{k=2}^{n}\frac{1}{c_n+n-k}\frac{1}{\log k},$$

where $c_n = 2n^x/(\sigma^2 s)$. For fixed, α, such that $0 < \alpha < 1$, write

$$\sum_{k=2}^{n}\frac{1}{c_n+n-k}\frac{1}{\log k} = \sum_{2\leq k\leq n^\alpha}(\cdot) + \sum_{n^\alpha\leq k\leq n}(\cdot)$$

Clearly, the first sum in the right-hand side of the equation vanishes as $n \to \infty$ because

$$\sum_{2\leq k\leq n^\alpha}\frac{1}{c_n+n-k}\frac{1}{\log k} \leq \frac{n^\alpha}{n-n^\alpha} \to 0 \qquad \text{as } n \to \infty.$$

For the second sum, we have

$$\frac{1}{\log n}\sum_{n^\alpha\leq k\leq n}\frac{1}{c_n+n-k} \leq \sum_{n^\alpha\leq k\leq n}\frac{1}{c_n+n-k}\frac{1}{\log k}$$

$$\leq \frac{1}{\alpha\log n}\sum_{n^\alpha\leq k\leq n}\frac{1}{c_n+n-k}$$

and passing to the limit

$$\lim_{n\to\infty}\frac{1}{\log n}\sum_{n^\alpha\le k\le n}\frac{1}{c_n+n-k}=\lim_{n\to\infty}\frac{1}{\log n}\sum_{k=0}^{n}\frac{1}{c_n+n-k}$$

$$=\lim_{n\to\infty}\frac{1}{\log n}\int_0^n\frac{1}{c_n+u}\,du$$

$$=\lim_{n\to\infty}\frac{1}{\log n}\left[\log\left(1+\frac{n}{c_n}\right)\right]$$

$$=\lim_{n\to\infty}\frac{1}{\log n}\left[\log\left(1+\frac{\sigma^2 s}{2}n^{1-x}\right)\right]$$

$$=1-x.$$

Therefore,

$$1-x\le\liminf_{n\to\infty}\sum_{k=2}^{n}\frac{1}{c_n+n-k}\,\frac{1}{\log k}$$

$$\le\limsup_{n\to\infty}\sum_{k=2}^{n}\frac{1}{c_n+n-k}\,\frac{1}{\log k}\le\frac{1-x}{\alpha}.$$

Since $\alpha\in(0,1)$ is arbitrary, it follows that

$$\lim_{n\to\infty}\sum_{k=2}^{n}\frac{1}{c_n+n-k}\,\frac{1}{\log k}=1-x.$$

Combining this with [1.55], we obtain for $s>0$ and $0<x<1$

$$\lim_{n\to\infty}E\left[\exp\left\{-s\frac{Y_n^0}{n^x}\right\}\right]=x.$$

This limit is the Laplace transform of the degenerate distribution with mass x at 0 and mass $1-x$ at ∞. Applying the continuity theorem for Laplace transforms (see theorem A2.4 in Appendix 2), we have for any $y>0$

$$\lim_{n\to\infty}P\left(\frac{Y_n^0}{n^x}\le y\right)=x. \qquad [1.56]$$

Setting $y=1$ and taking logarithms of both sides of the inequality in [1.56] completes the proof of the theorem. ∎

1.3.2. *Critical process with decreasing to zero immigration*

Similarly to section 1.2.2, let $\{Y_n^0\}_{n\geq 0}$ be a critical FP process with **decreasing to zero (vanishing) immigration**. That is, $Y_0^0 = 0$ and for $n = 0, 1, \ldots$

$$Y_{n+1}^0 = \sum_{i=1}^{Y_n^0} X_{n,i} + I_{n+1}\mathcal{I}\{Y_n^0 = 0\},$$

(with $\sum_1^0 = 0$), where $\{I_n\}_{n\geq 1}$ are not identically distributed and $E[I_n] \to 0$ as $n \to \infty$. Recall that $g_n(s) = E[s^{I_n}]$, $a_n = g_n'(1)$, and $c_n = g_n''(1)$. Suppose, similarly to section 1.2.2, that $\sup_n a_n < \infty$ and $\sup_n c_n < \infty$. As one can expect, the faster the convergence of a_n to zero is, the less influence the immigration has over the evolution of the population. Indeed, under some conditions on c_n, if $a_n \sim n^{-r}L(n)$ as $n \to \infty$, where $0 \leq r < 1$ and $L(n)$ is an s.v.f. ("slowly" vanishing immigration), then the limit [1.52] for a FP process with time-independent immigration still holds. If $\sum_{n=0}^{\infty} a_n < \infty$ ("fast" vanishing immigration), then in the long run the process behaves as an ordinary GW process and the limit [1.13] holds. However, a proper "tuning" of the rate in which a_n approaches zero yields new limiting results. As we will see in theorem 1.14, it is possible for the process $\{Y_n^0\}_{n\geq 0}$ to have two different types of non-degenerate trajectories.

THEOREM 1.14.– *Let $\{Y_n^0\}_{n\geq 0}$ be a critical FP process with $\sigma^2 < \infty$ and time-dependent immigration component $\{I_n\}_{n\geq 0}$. Assume $a_n \sim n^{-1}L(n)$, where $L(n)$ is an s.v.f. at infinity and $c_n = o(a_n \log n)$ as $n \to \infty$.*

1) If the s.v.f. $L^*(n) := \sum_{k=1}^{\infty} a_k$ tends to infinity, then the probability of not-hitting zero at time n satisfies

$$P\left(Y_n^0 > 0\right) \sim \frac{2}{\sigma^2 n}(L^*(n) + L(n)\log n) \qquad [1.57]$$

and the first two moments of the process tend to infinity as follows

$$E\left[Y_n^0\right] \sim L^*(n), \quad E\left[(Y_n^0)^2\right] \sim \sigma^2 n L^*(n).$$

2) If, in addition,

$$\lim_{n\to\infty} \frac{L(n)\log n}{L^*(n)} = \theta, \quad 0 < \theta < \infty, \qquad [1.58]$$

then for $0 \leq x \leq 1$

$$\lim_{n\to\infty} P\left(\frac{\log Y_n^0}{\log n} \leq x \,\middle|\, Y_n^0 > 0\right) = \frac{\theta}{\theta+1}x =: G_1(x) \qquad [1.59]$$

while for $x \geq 0$

$$\lim_{n\to\infty} P\left(\frac{2Y_n^0}{\sigma^2 n} \leq x \,\middle|\, Y_n^0 > 0\right) = \frac{\theta}{\theta+1} + \frac{1}{\theta+1}\left(1 - e^{-x}\right) =: G_2(x). \quad [1.60]$$

COMMENT.– (i) Since $\lim_{x\to 1} G_1(x) = \lim_{x\to 0} G_2(x)$, the limits in [1.59] and [1.60] represent two types of non-degenerate trajectories of $\{Y_n^0\}_{n\geq 0}$. Let ζ_1 have a Uniform (0,1) distribution and ζ_2 have an Exponential $(\sigma^2/2)$ distribution. Then as $n \to \infty$

(A) $Y_n^0 \sim n^{\zeta_1}$ with probability $\theta/(\theta+1)$;

(B) $Y_n^0 \sim \zeta_2 n$ with probability $1/(\theta+1)$.

It turns out that other branching models, some of which will be considered in the following chapters (see also [VAT 79]), obey such asymptotic behavior.

(ii) The assumption [1.58] holds, for example, if $L(n) \sim (\log n)^r L_1(\log n)$ for $r > -1$, where $L_1(n)$ is an s.v.f. It can be shown that, in this case, $\theta = r + 1$.

(iii) It is interesting to compare theorem 1.14 with theorem 1.10. If in [1.37] $\alpha = 1$ and $L_f(x) \to c < \infty$ as $x \to 1$ (finite offspring variance case), then $L(x) \sim c^{-1} \log x$ and $(1 - f_n(0)) \sim 2n/\sigma^2$. In this case, it can be verified that the limit results [1.39] and [1.40] for the GWI process coincide with the limit results [1.59] and [1.60] for the FP process. Therefore, the reason for the occurrence of new limit distributions is the rate immigration converges to zero rather than whether the immigration is unrestricted or at zero only. In fact, the only difference between GWI process and FP process is that in the latter immigration is not allowed during the life-period of the process. However, if the immigration approaches zero fast enough, then (due to the criticality of the underlying process) the duration of the life-period tends to zero, which implies the equivalent asymptotic behavior of the two processes.

Proof. (For a more detailed proof, see [MIT 84b].) We will outline the proof of part 2 only. Observe as before, that for $0 \leq s \leq 1$

$$E\left[s^{Y_n^0} \mid Y_n^0 > 0\right] = 1 - \frac{1 - H_n^0(s)}{P\left(Y_n^0 > 0\right)}. \qquad [1.61]$$

Similarly to [1.46], we have

$$H_n^0(s) = 1 - \sum_{k=0}^{n-1}(1 - g_k(f_{n-k}(s))H_k^0(0). \qquad [1.62]$$

Note that the criticality of the process implies that for any $\varepsilon > 0$, there exists N, such that for $n \geq N$

$$\frac{1-\varepsilon}{(1-s)^{-1} + 2n/\sigma^2} \leq 1 - f_n(s) \leq \frac{1+\varepsilon}{(1-s)^{-1} + 2n/\sigma^2}, \quad 0 \leq s \leq 1 \quad [1.63]$$

and, by the properties of s.v.f.s,

$$(1-\varepsilon)n^{-1}L(n) \leq a_n \leq (1+\varepsilon)n^{-1}L(n). \qquad [1.64]$$

Therefore,

$$(1-\varepsilon)^2 \frac{S_n(s_n)}{P\left(Y_n^0 > 0\right)} \leq \sum_{k=0}^{n} a_k \frac{1 - f_{n-k}(s_n)}{P\left(Y_n^0 > 0\right)} \leq (1+\varepsilon)^2 \frac{S_n(s_n)}{P\left(Y_n^0 > 0\right)}, \quad [1.65]$$

where

$$S_n(s_n) := \sum_{k=0}^{n-1} \frac{L(k)}{k[(1-s_n)^{-1} + 2(n-k)/\sigma^2]}. \qquad [1.66]$$

Consider [1.59]. Setting $s_n = \exp\{-u/n^x\}$ in [1.66], one can see that for any $\varepsilon > 0$ there exists N_1, such that for any fixed N satisfying $N_1 \leq N \leq n - N_1$ we have

$$\begin{aligned} S_n\left(e^{-u/n^x}\right) &= \sum_{k=N}^{n-N} \frac{L(k)}{k[n^x/u + 2(n-k)/\sigma^2]} \\ &\sim \frac{2}{\sigma^2 n}[L^*(n) + (1-x)L(n)\log n]. \end{aligned} \qquad [1.67]$$

Referring to [1.57], [1.65], and the Sandwich theorem, we obtain

$$\lim_{n\to\infty} \frac{1}{P\left(Y_n^0 > 0\right)} \sum_{k=0}^{n} a_k \left(1 - f_{n-k}(s_n)\right) = \frac{1 + (1-x)\theta}{\theta + 1}. \qquad [1.68]$$

On the other hand, the Taylor expansion

$$1 - g_n(s) = a_n(1-s) - \bar{c}_n(s)(1-s)^2/2$$

holds, where under the assumptions of the theorem, $\bar{c}_n(s) \sim c_n$ as $n \to \infty$ and $s \uparrow 1$. It can be shown that as $n \to \infty$

$$\sum_{k=0}^{n-1} (1 - g_k(f_{n-k}(s_n))\, H_k^0(0) = \sum_{k=0}^{n-1} a_k \left(1 - f_{n-k}(s_n)\right) H_k^0(0) \qquad [1.69]$$

$$+\frac{1}{2} \sum_{k=0}^{n-1} \bar{c}_n(s_n) \left(1 - f_{n-k}(s_n)\right)^2 H_k^0(0)$$

$$\sim \sum_{k=0}^{n} a_k \left(1 - f_{n-k}(s_n)\right).$$

Taking into account [1.61]–[1.62] and [1.68]–[1.69], we obtain

$$\lim_{n\to\infty} E\left[\exp\left\{-\frac{uY_n^0}{n^x}\right\} \,\middle|\, Y_n^0 > 0\right] = \frac{\theta}{\theta+1} x.$$

Observe that this limit is the Laplace transform of the degenerate distribution with mass $\theta x/(\theta + 1)$ at 0 and mass $1 - \theta x/(\theta + 1)$ at ∞. Applying the continuity theorem for Laplace transforms, we have for any $y > 0$

$$\lim_{n\to\infty} P\left(\frac{Y_n^0}{n^x} \le y\right) = x. \qquad [1.70]$$

Setting $y = 1$ and taking logarithms of both sides of the inequality in [1.70] completes the proof of [1.59].

To prove [1.60], let us substitute $\exp\{-2u/\sigma^2 n\}$ for s_n in [1.66]. One can obtain for any $u > 0$ as $n \to \infty$

$$S_n(e^{-2u/\sigma^2 n}) \sim \frac{2u}{\sigma^2(1+u)} \frac{L^*(n)}{n}. \qquad [1.71]$$

Referring to [1.57], [1.65], and the Sandwich theorem, we have

$$\lim_{n\to\infty} \frac{1}{P(Y_n^0 > 0)} \sum_{k=0}^{n} a_k(1 - f_{n-k}(s_n)) = \frac{1}{\theta+1}\frac{u}{u+1}. \qquad [1.72]$$

Next, repeating the arguments in the proof of [1.59], we obtain for $u > 0$

$$\lim_{n\to\infty} E\left[\exp\left\{-\frac{2uY_n^0}{\sigma^2 n}\right\} \,\middle|\, Y_n^0 > 0\right] = \frac{\theta}{\theta+1} + \frac{1}{\theta+1}\frac{1}{1+u}.$$

Now, [1.60] follows from the continuity theorem for Laplace transforms. ∎

1.4. Background and bibliographical notes

The first limit theorems for the ordinary GW process were proved by Kolmogorov [KOL 38] in 1938 (see [SEN 69]), while studying the asymptotic behavior of the probability of non-extinction (survival) by time n, namely, $P(T > n)$ as $n \to \infty$. Kolmogorov also coined the term "branching process". As Sevastyanov affirmed in [SEV 09, p. 88], the word "vetvyashchiisya" (branching) did not exist in the Russian language before Kolmogorov coined it. This marks the beginning of the modern era in the study of branching processes and their diverse applications. The first monograph treatment is given in [HAR 63]. An excellent historical review concerning this period of the theory and applications of branching processes is included in [KEN 66]. A more recent addition to the historical account is given in [ULA 90].

The GW process with immigration seems to have first appeared (see [SEN 69]) within a genetic context in [HAL 49] describing the development of mutant genes in a large population. The "immigration" component was because of recurrent mutation at each generation. Seminal results, including a Gamma limiting theorem for the continuous time counterpart of the GWI process, were proved in [SEV 57]. During the period 1960–1990, branching processes with different regimes of immigration were studied intensively by researchers from Australia, Bulgaria (see [MIT 09]), and Uzbekistan among others. The classical GWI process is well understood nowadays. However, to cite [PAK 06] here: "The literature on immigration–branching processes is widely scattered and there is no single source that can put the reader in touch with even a substantial portion of it". Early versions of its limit theory can be found in the monographs [SEV 71, ATH 72, JAG 75], and more complete results are given in [PAK 79] and [ASM 83]. A review of results for GWI processes is given in [VAT 93]. For a more general model with reproduction-dependent immigration, we refer to [RAH 95]. It is not our intention to present a complete account of these investigations. Rather, selected results regarding models with immigration, relevant to the material in the rest of the book, are included.

The exposition in section 1.1 closely follows [VAT 08]. The proof of the uniqueness in theorem 1.4 is from [ATH 72]. For a contemporary treatment of GW processes utilizing size-based GW trees (including a proof of Kesten–Stigum theorem and its refinement), we refer to [LYO 16, Chapter 12]. In section 1.2, the classification [1.26] of the Markov chain $\{Y_n\}_{n\geq 0}$ summarizes results from [HEA 66, SEN 68, PAK 71c, PAK 71b] and [ZUB 72]. Additional results were obtained in [PAK 75]. The decomposition [1.23] is taken from [SEN 70]. Theorem 1.7 and the following comment are from [HEA 66]. See also [PAK 71b] for more information about the limiting–stationary distribution. Theorem 1.8 is demonstrated in [PAK 74]. The proof of theorem 1.9 is taken from [VAT 08]. Let us point out here a more general method for proving limit theorems for branching processes given in [BOR 88]. The interpretation of theorem 1.9 for the critical GWI process is from [ATH 72] and [JAG 75]. Theorem 1.10 is from [RAH 86]. Continuous time Markov branching processes with decreasing time–dependent immigration are first considered in [BRU 78]. In section 1.3, the proof of theorem 1.13 for the critical FP is an elaboration of the original proof in [FOS 71]. Theorems 1.11 and 1.12 are obtained in [PAK 74]. The material for FP process with time-dependent immigration is from [MIT 84b] (see also [MIT 84a] for the continuous time case). Note that it is shown in [RAH 87] that one can study the FP processes with decreasing immigration, comparing them to GWI processes with a specific decreasing immigration rate, not dependent on the state of the process. This is reflected in Comment (iii) after theorem 1.14.

2

Branching Processes with Migration

"The real voyage of discovery consists not in seeking new landscapes, but in seeking new eyes".
– Marcel Proust

In this chapter, we study Galton–Watson processes with migration (GWM processes). These processes are controlled branching processes, which include as particular cases models with different regimes of immigration and emigration. We will present results about (i) life-periods, (ii) probability of hitting zero, (iii) moments, and (iv) limit theorems for the population size, and will make comparison with processes with immigration. Essential techniques in the proofs include asymptotic analysis of equations for probability generation functions and results about Markov chains and renewal processes.

2.1. Galton–Watson process with migration

Let the random variables $\{X_{n,i} : n = 0, 1, \ldots, i = 1, 2, \ldots\}$ be the offspring variables as defined in Chapter 1 with common distribution $\{p_k\}_{k\geq 0}$. Let us define a set $\{I_n\}_{n\geq 0}$ of non-negative integer–valued i.i.d. (immigration) random variables on the same probability space, independent from the offspring variables and with common distribution $\{g_k\}_{k\geq 0}$.

DEFINITION 2.1.– *The discrete time homogeneous Markov chain $\{Y_n\}_{n\geq 0}$ is called a GWM process if $Y_0 = 0$ and for $n = 0, 1, \ldots$*

$$Y_{n+1} = \sum_{i=1}^{Y_n} X_{n,i} + M_{n+1}, \qquad [2.1]$$

(with $\sum_1^0 = 0$*), where for* $p + q + r = 1$

$$M_{n+1} = \begin{cases} -X_{n,1}\mathcal{I}\{Y_n > 0\} & \textit{with probability } p, \quad \textit{(emigration)} \\ 0 & \textit{with probability } q, \quad \textit{(no migration)} \\ I_{n+1} & \textit{with probability } r, \quad \textit{(immigration)} \end{cases} \qquad [2.2]$$

is the migration component.

We will assume without further comment that p and r are positive, and p_0, $p_0 + p_1$ and g_0 are all positive and less than one. Throughout this chapter, we will assume finite offspring variance σ^2 and finite immigration mean a. As in the previous chapter the offspring mean will be denoted by m.

The process $\{Y_n\}_{n\geq 0}$ can be interpreted as follows. Three scenarios are possible (see Figure 2.1): (i) the offspring of one individual is removed (emigration) with probability p, (ii) there is no migration with probability q, or (iii) I_n individuals join the population (immigration) with probability r. Thus, the state zero is a reflecting barrier for $\{Y_n\}_{n\geq 0}$. The emigration in this model can be regarded as "reversed" (negative) immigration since the branching process is modified to allow both positive and negative increments.

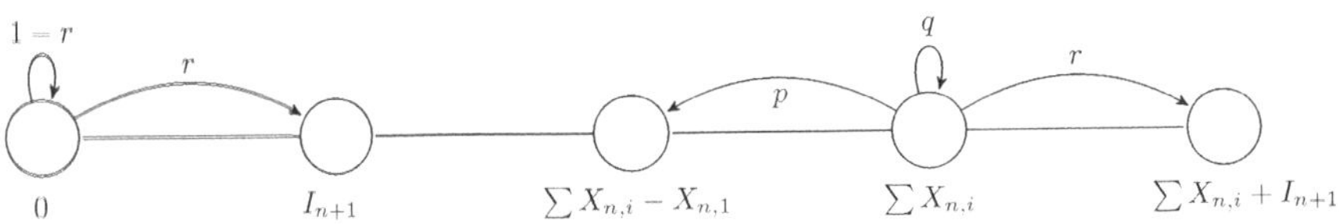

Figure 2.1. *Diagram for a GWM process*

Although the terminology we use here is motivated by modeling biological systems, it is worth mentioning the connection with systems in the fields of quantum optics and atomic physics. The state of these systems is often described in terms of emission and absorption of a number of energy quanta. Thus, births are the analog of stimulated emission. The emigration and immigration are the analogs of absorption and spontaneous emission, respectively, see [HOP 14, p. 32].

As far as Galton–Watson processes with emigration (catastrophes, disasters) are concerned, under the fundamental assumption of independence of the individual evolutions, several possible scenarios are studied. The first is **independent emigration**, in which each individual may choose to leave the population with a given fixed probability. In this case, emigration can be considered as death of the individual resulting in a simple modification of the offspring distribution. If q is the probability of emigration and p_k the probability of having k children, the probability of having k non-emigrating children is (e.g. [HAC 05, p. 54])

$$\sum_{j \geq k} \binom{j}{k} (1-q)^k q^{j-k} p_j.$$

Therefore, the process with individual emigration is an ordinary GW process and does not require a separate analysis. A second possibility is that emigration chance depends on the total number of newborn individuals or on the size of the adult population. Such **population-size-dependent emigration** can be accommodated within the so-called population-size-dependent branching processes or with so-called GW processes with "killing" (see [PAK 84] and [SAG 17]). More difficult for analysis, but probably also more interesting, are cases in which emigration decisions of different individuals within one generation are not independent. Unfortunately, as yet there are no results on the dynamics of such processes.

To study GWM processes, we need the p.g.f. $\Phi_n(s) := E\left[s^{Y_n}\right], 0 \leq s \leq 1$, and the migration regime function

$$d(s) := \frac{p}{f(s)} + q + rg(s),$$

where, recall from Chapter 1, $f(s)$ and $g(s)$ are the p.g.f.s of the number of offspring and immigrants, respectively. Using the definition of $\{Y_n\}_{n \geq 0}$, we obtain the following equation for $0 \leq s \leq 1$

$$\Phi_{n+1}(s) = \sum_{k=0}^{\infty} P(Y_n = k) E\left[s^{Y_{n+1}} \mid Y_n = k\right] \qquad [2.3]$$

$$= \sum_{k=1}^{\infty} P(Y_n = k) f^k(s) \left(\frac{p}{f(s)} + q + rg(s)\right) + P(Y_n = 0)(1 - r + rg(s))$$

$$= d(s)\Phi_n(f(s)) - p\Phi_n(0)\left(\frac{1}{f(s)} - 1\right).$$

Iterating [2.3], we have for $0 \leq s \leq 1$

$$\Phi_n(s) = d_n(s) - p \sum_{k=0}^{n-1} \Phi_{n-1-k}(0) \left(\frac{1}{f_{k+1}(s)} - 1 \right) d_k(s), \qquad [2.4]$$

where $d_n(s) := \prod_{k=0}^{n-1} d(f_k(s))$ for $n = 1, 2, \ldots; d_0(s) = 1$.

Let us introduce a measure of migration balance $\theta \in \mathbf{R}$ defined by

$$\theta := \frac{2E[M_n^+]}{Var[X_{0,1}]} = \frac{2(ra - pm)}{\sigma^2},$$

where M_n^+ is the migration outside zero. If $\theta > 0$, then immigration dominates over emigration; if $\theta = 0$, then immigration levels emigration; and if $\theta < 0$, then emigration dominates over immigration.

Differentiating [2.3] at $s = 1$ and investigating the asymptotic behavior of $\Phi_n(0)$, it can be obtained that as $n \to \infty$

$$E[Y_n] \sim \begin{cases} c_1 m^n & \text{if} \quad m > 1 \\ c_2 n & \text{if} \quad m = 1, \quad \theta > 0 \\ c_3 n / \log n & \text{if} \quad m = 1, \quad \theta = 0 \\ c_4 n^{1+\theta} & \text{if} \quad m = 1, \quad \theta < 0 \\ c_5 & \text{if} \quad m < 1, \end{cases} \qquad [2.5]$$

where $c_i := c_i(\theta)$ for $1 \leq i \leq 5$ are certain positive constants for fixed θ.

DEFINITION 2.2.– *The process $\{Y_n\}_{n \geq 0}$ is said to be subcritical, critical or supercritical if $m < 1$, $m = 1$ or $m > 1$, respectively.*

The parameter θ plays a role similar to that of γ for the GWI process from Chapter 1. Assuming $\sigma^2 < \infty$ and $a < \infty$, the aperiodic and irreducible Markov chain $\{Y_n\}_{n \geq 0}$ is classified as

$$\begin{cases} \text{transient} & \text{if} \quad m > 1 \\ \text{transient} & \text{if} \quad m = 1, \quad \theta \geq 1 \\ \text{null–recurrent} & \text{if} \quad m = 1, \quad \theta = 1 \quad \text{and} \quad [2.7] \\ \text{null–recurrent} & \text{if} \quad m = 1, \quad 0 \leq \theta < 1 \\ \text{positive recurrent} & \text{if} \quad m = 1, \quad \theta < 0 \\ \text{positive recurrent} & \text{if} \quad m < 1, \end{cases} \qquad [2.6]$$

where

$$E\left[X_{0,1}^2 \log(1+X_{0,1})\right] < \infty \quad \text{and} \quad \sum_{k=2}^{\infty}\left(\sum_{j=0}^{k} p_j g_{k-j}\right) k \log k < \infty. \quad [2.7]$$

The duration τ of a **life-period** of a GWM process is defined as the time between two consecutive visits to zero. In the **critical case** $(m = 1)$, it can be proven (under some additional moment conditions) that as $n \to \infty$

$$P(\tau > n) \sim \begin{cases} c_1 & \text{for} \quad \theta > 1 \\ c_2/\log n & \text{for} \quad \theta = 1 \\ c_3/n^{1-\theta} & \text{for} \quad \theta < 1, \end{cases} \quad [2.8]$$

where $c_i := c_i(\theta)$ for $1 \le i \le 3$ are positive constants. Notice that, unlike γ in a GWI process, the parameter θ takes on negative values when the emigration dominates over the immigration on average. In the **subcritical case**, it can be proven that

$$E[\tau] = \frac{1}{(1-g_0)d}\sum_{k=0}^{\infty} d_k(0)(1 - g(f_k(0))),$$

where $d = \lim_{n\to\infty} d_n(0)$.

2.2. Limit theorems

In this section, we discuss limit theorems for GWM processes with non negative and negative balances between immigration and emigration. If the migration mean, $E[M_n^+]$, is non-negative (emigration does not exceed immigration on average), a comparison is made with similar results for GWI processes from Chapter 1. If the migration mean is negative (emigration exceeds immigration on average), we contrast the limiting behavior of these two sub-classes of controlled branching processes.

2.2.1. *Non-critical processes*

In the **subcritical case**, GWM and GWI processes have similar limiting behavior. More precisely, the following theorem holds.

THEOREM 2.1.– *Let $\{Y_n\}_{n\geq 0}$ be a subcritical GWM process with $\sigma^2 < \infty$ and $a < \infty$. There exists a limiting–stationary distribution $\{\pi_k\}_{k\geq 0}$ given by*

$$\sum_{k=0}^{\infty} \pi_k s^k = d_\infty(s) - p\pi_0 \sum_{k=0}^{\infty} d_k(s) \left(\frac{1}{f_{k+1}(s)} - 1 \right), \qquad 0 \leq s \leq 1, \qquad [2.9]$$

where $d_\infty(s) := \prod_{k=0}^{\infty} d(f_k(s))$ is convergent.

Proof. We begin with proving that $\{Y_n\}_{n\geq 0}$ is an irreducible and aperiodic Markov chain. We denote $p_{ij}^{(n)} := P(Y_{k+n} = j \mid Y_k = i)$ for $k \geq 0$ and $n \geq 1$. Then the one-step probability for returning to 0 is $p_{00}^{(1)} = 1 - r(1 - g_0) > 0$. Hence, the state 0 has period 1. All states of $\{Y_n\}_{n\geq 0}$ are accessible from 0, that is, for every state j there exists $n_j > 0$ such that $p_{0j}^{(n_j)} > 0$. Furthermore, for every two states i and j, we have $p_{ij}^{(n_j+1)} \geq p_{i0}^{(1)} p_{0j}^{(n_j)} > 0$ because, by the definition of the process,

$$p_{i0}^{(1)} = p p_0^{\max(i-1,0)} + q p_0^i + r p_0^i g_0 > 0.$$

Therefore, $\{Y_n\}_{n\geq 0}$ is irreducible and aperiodic. Furthermore, it can be proved (see [YAN 84c]) that $\lim_{n\to\infty} p_{00}^{(n)} > 0$. Thus, by the theory of Markov chains (see [FEL 66, p. 356]), there exists a unique stationary distribution $\{\pi_k\}_{k\geq 0}$, say, such that $0 \leq s \leq 1$

$$\lim_{n\to\infty} \Phi_n(s) = \Pi(s), \qquad \Pi(s) := \sum_{k=0}^{\infty} \pi_k s^k.$$

Letting $n \to \infty$ in [2.3], we see that $\Pi(s)$ satisfies the relation

$$\Pi(s) = d(s)\Pi(f(s)) - p\pi_0 \left(\frac{1}{f(s)} - 1 \right), \qquad 0 < s \leq 1. \qquad [2.10]$$

Iterating [2.10], it is not difficult to prove by induction that

$$\Pi(s) = d_n(s)\Pi(f_n(s)) - p\pi_0 \sum_{k=0}^{n-1} d_k(s) \left(\frac{1}{f_{k+1}(s)} - 1 \right). \qquad [2.11]$$

Observe that for $0 \leq s \leq 1$

$$0 < p_0 \leq f_k(0) \leq |f_k(s)| \leq 1, \quad |1 - f_k(s)| \leq m^k |1 - s| \leq 2m^k,$$

and

$$|1 - g(f_k(s))| \leq a|1 - f_k(s)| \leq 2am^k.$$

Therefore,

$$|d(f_k(s)) - 1| = \left| \frac{p}{f_{k+1}(s)} + q + rg(f_k(s)) - 1 \right| \leq 2m^k \left(\frac{mp}{f(0)} + ra \right).$$

The above inequalities imply that uniformly for $0 \leq s \leq 1$

$$\lim_{n \to \infty} d_n(s) =: d_\infty(s) \quad \text{and} \quad \sum_{k=0}^{\infty} d_k(s) \left(\frac{1}{f_{k+1}(s)} - 1 \right) < \infty.$$

The last equations justify taking the limit as $n \to \infty$ in [2.11] that yields [2.9]. ∎

The next result (for its proof see [DYA 96]) sheds light on the limiting stationary distribution in theorem 2.1. It is a limit theorem for the so-called "close-to-critical" process when $m \uparrow 1$. Define the class $\mathcal{K} = \{f(s) : 0 < s < 1\}$ of p.g.f.s. as follows. For every $f \in \mathcal{K}$, there exists $h(s) = o(s^2)$ as $s \to 0$, such that

$$f(1-s) = 1 - ms + cs^2 + \alpha(s), \qquad f(0) > 0, \qquad [2.12]$$

where $\alpha(s) \leq h(s)$, $0 < m < 1$, and c is a positive constant.

THEOREM 2.2.– *Let $\{Y_n\}_{n\geq 0}$ be a GWM process with offspring p.g.f. $f(\cdot)$ belonging to the class $\mathcal{K}$. Let Y be the limiting random variable in theorem 2.1. If $\theta = 0$ and $Var[I_1] < \infty$, then*

$$\lim_{m \uparrow 1} P\left(\frac{\log Y}{\log(1/(1-m))} \leq x \right) = x, \qquad x \in (0,1).$$

COMMENT.– For a limit theorem of this kind for processes with immigration and a related discussion, we refer to [SEN 69].

As it could be expected, in **the supercritical case**, the migration does not have a significant effect on the long-term evolution of the population. It can be proved (see [KHA 80]) that $\{Y_n/m^n\}_{n\geq 0}$ converges a.s. to a non-negative random variable V with $E[V] < \infty$. Furthermore, $P(V = 0) = 1$ if and only if $E\left[X_{0,1} \log(1 + X_{0,1})\right] = \infty$.

2.2.2. *Critical processes with non-negative migration mean*

In this section and section 2.2.3, we will investigate separately the two cases: $\theta \geq 0$ and $\theta < 0$. First, consider a critical GWM process with $\theta \geq 0$, that is, on average at each generation, **the emigration does not exceed the immigration**. Under this condition, the study of GWM processes is analytically more involved than that for GWI processes. However, the final results show that such GWM processes exhibit behavior, which is qualitatively similar to that of GWI processes. To illustrate the similarities, we formulate a series of limit theorems which are counterparts of those for GWI processes discussed in Chapter 1.

THEOREM 2.3.– *Let* $\{Y_n\}_{n\geq 0}$ *be a critical GWM process with* $\sigma^2 < \infty$ *and* $a < \infty$.

1) If $\theta > 0$ *(immigration dominates over emigration), then*

$$\lim_{n\to\infty} P\left(\frac{2Y_n}{\sigma^2 n} \leq y\right) = P(Y \leq y), \qquad y > 0, \tag{2.13}$$

where the limit Y *has Gamma distribution with parameters* θ *and* 1.

2) If $\theta = 0$ *(immigration equals emigration on average), then*

$$\lim_{n\to\infty} P\left(\frac{\log Y_n}{\log n} \leq y\right) = P(Y \leq y), \qquad 0<y<1, \tag{2.14}$$

where the limit Y *has Uniform distribution over the interval* $(0, 1)$.

COMMENT.– The results [2.13] and [2.14] are analogs of those for GWI (theorem 1.9) and FP processes (theorem 1.13) in Chapter 1, respectively.

Next, let us modify the GWM processes defined by [2.1], allowing **time-dependent migration** as follows.

DEFINITION 2.3.– *The process* $\{\tilde{Y}_n\}_{n\geq 0}$ *is called a branching process with time-dependent migration if it is defined by* [2.1], *with the controlling migration parameters* $p = p_n$, $q = q_n$ *and* $r = r_n$ *depending on time.*

Note that $\{\tilde{Y}_n\}_{n\geq 0}$ is a non-homogeneous Markov chain. Until the end of this section, we assume

$$\theta_n := \frac{2(r_n a - p_n)}{\sigma^2} = 0,$$

that is, at each generation **the emigration equals the immigration** on average. In addition, we let both controlling parameters p_n and q_n decrease to zero at the same rate. Under these assumptions, a key role in the limiting behavior of the process is played by the series $\sum_{n=0}^{\infty} p_n$ (or, equivalently, by $\sum_{n=0}^{\infty} r_n$).

THEOREM 2.4.– *Let $\{\tilde{Y}_n\}_{n\geq 0}$ be a critical GWM process with time-dependent migration and $\theta_n = 0$. Suppose one of the following two conditions holds as $n \to \infty$*

(i) $p_n \sim L(n)/n^v$ for $0 < v < 1$ where $L(n)$ is an s.v.f. at infinity;

(ii) $p_n = O\left(\frac{1}{\log n}\right)$.

Then

$$\lim_{n\to\infty} P\left(\frac{\log \tilde{Y}_n}{\log n} \leq x \middle| \tilde{Y}_n > 0\right) = P\left(\tilde{Y} \leq x\right), \qquad 0 < x < 1$$

where the limit $\tilde{Y}$ has Uniform distribution over the interval $(0, 1)$.

COMMENT.– Recall that a uniformly distributed over (0,1) limit was obtained for the FP process in theorem 1.13, Chapter 1.

THEOREM 2.5.– *Let $\{\tilde{Y}_n\}_{n\geq 0}$ be a critical GWM process with time-dependent migration and $\theta_n = 0$. Assume that as $n \to \infty$*

(i) $p_n \sim L(n)/n$, where $L(n)$ is an s.v.f. at infinity;

(ii) $\lim_{n\to\infty} \frac{p_n n \log n}{\sum_{k=1}^{n} p_k} = c$ for $0 \leq c < \infty$.

Then, for $0 < x < 1$

$$\lim_{n\to\infty} P\left(\frac{\log \tilde{Y}_n}{\log n} \leq x \middle| \tilde{Y}_n > 0\right) = \frac{c}{1+c} x, \qquad [2.15]$$

while for $x > 0$

$$\lim_{n\to\infty} P\left(\frac{2\tilde{Y}_n}{\sigma^2 n} \leq x \middle| \tilde{Y}_n > 0\right) = \frac{c}{1+c} + \frac{1}{1+c}\left(1 - e^{-x}\right). \qquad [2.16]$$

COMMENT.– Theorem 2.5 is an analog of theorem 1.14, Chapter 1 for the FP process with immigration decreasing to zero.

THEOREM 2.6.– *Let $\{\tilde{Y}_n\}_{n\geq 0}$ be a critical GWM process with time–dependent migration and $\theta_n = 0$. If $\sum_{k=1}^{\infty} p_k < \infty$, then for $x > 0$*

$$\lim_{n\to\infty} P\left(\frac{2\tilde{Y}_n}{\sigma^2 n} \leq x \middle| \tilde{Y}_n > 0\right) = 1 - e^{-x}.$$

COMMENT.– Theorem 2.6 is an analog of the classical Kolmogorov–Yaglom result for the ordinary GW process (theorem 1.5 in Chapter 1). It turns out that the convergence of $\sum_{n=0}^{\infty} p_n$ and $\sum_{n=0}^{\infty} r_n$ ensures that the migration disappears without a trace so fast that the GWM process with non-homogeneous migration has the same asymptotic behavior as the ordinary GW process.

2.2.3. *Critical processes with negative migration mean*

Let us consider critical GWM processes with $\theta < 0$, that is, at each generation **the emigration exceeds the immigration** on average. We will contrast the limit results for GWM processes with those for GWI processes.

We need the following moment assumptions:

$$\begin{aligned} E\left[X_{0,1}^2 \log X_{0,1}\right] &< \infty, \quad E\left[I_0^{1-\theta}\right] < \infty \text{ if } -1 < \theta < 0 \\ E\left[X_{0,1}^2 \log^2 X_{0,1}\right] &< \infty, \quad E\left[I_0^2 \log^2 I_0\right] < \infty \text{ if } \quad \theta = -1 \\ E\left[X_{0,1}^{1-\theta}\right] &< \infty, \quad E\left[I_0^{1-\theta}\right] < \infty \text{ if } \quad \theta < -1. \end{aligned} \qquad [2.17]$$

In the proof of theorem 2.7 below, we will use the following lemma 2.1. For the lemma's proof, we refer to [YAN 04].

LEMMA 2.1.– *Let $\{Y_n\}_{n\geq 0}$ be a critical GWM process with $\sigma^2 < \infty$, $a < \infty$ and $\theta < 0$. Assume also that* [2.17] *hold. Then for $0 \leq s < 1$ and $n \to \infty$*

$$\frac{1}{d_n(s)} \sum_{k=0}^{n} d_k(s)(1 - g(f_k(s))) = \frac{a}{p - ra} + \varphi_n(s)(1 + o(1)),$$

where

$$\varphi_n(s) = \begin{cases} c_\theta(s)/n & \text{if} \quad \theta < -1 \\ c_1(s) \log n / n & \text{if} \quad \theta = -1 \\ c_\theta(s) n^\theta & \text{if} \quad -1 < \theta < 0, \end{cases} \qquad [2.18]$$

and $c_\theta(s)$ and $c_1(s)$ are known functions of s, which are independent from n.

Now we are in a position to state and prove the theorem.

THEOREM 2.7.– *Let* $\{Y_n\}_{n\geq 0}$ *be a critical GWM process with* $\sigma^2 < \infty$, $a < \infty$ *and* $\theta < 0$. *Then there exists a limiting–stationary distribution* $\{\pi_k\}_{k\geq 0}$. *If, additionally, the assumptions* [2.17] *hold, then the limit's p.g.f.* $\Pi(s)$, $0 \leq s \leq 1$ *is the unique solution among p.g.f.s of the equation*

$$\Pi(s) = d(s)\Pi(f(s)) - p\pi_0\left(\frac{1}{f(s)} - 1\right), \qquad \pi_0 = 1 - \frac{ra}{p} \in (0,1). \qquad [2.19]$$

Also, as $s \uparrow 1$

$$1 - \Pi(s) \sim \begin{cases} c_1 \dfrac{1}{(1-s)^\theta} & \text{for } -1 < \theta < 0 \\ c_2(1-s)\log\dfrac{1}{1-s} & \text{for } \quad \theta = -1 \\ c_3(1-s) & \text{for } \quad \theta < -1, \end{cases} \qquad [2.20]$$

where $c_i := c_i(\theta)$ *for* $1 \leq i \leq 3$ *are positive constants.*

COMMENT.– For the critical GWI process, the existence of a limiting–stationary distribution is proved in [SEN 68] under the assumption of infinite offspring variance. In contrast, due to the migration regime with dominating emigration, theorem 2.7 holds when $\sigma^2 < \infty$.

Proof. First, we prove the existence of a limiting–stationary distribution satisfying [2.19]. Define the n-step transition probabilities from 0 to $k \geq 0$ without passing through 0 (a.k.a. taboo probability, see [CHU 67, p. 43]), by

$${}_0p_{0k}^{(n)} := P(Y_{n_0+n} = k,\ Y_{n_0+i} > 0,\ 1 \leq i < n \mid Y_{n_0} = 0),\ {}_0p_{0k}^{(0)} = 0.$$

If $u_k := P(\tau > k)$, where τ is the duration of a life period, then

$$P(\tau = n) = u_{n-1} - u_n = \frac{{}_0p_{00}^{(n+1)}}{1 - {}_0p_{00}(1)} = \frac{{}_0p_{00}^{(n+1)}}{r(1-g_0)}.$$

Since for $n = 1, 2, \ldots$

$$P(Y_n = 0) = \sum_{i=1}^{n} {}_0p_{00}^{(i)} P(Y_{n-i} = 0),$$

it can be obtained (see [YAN 04]) that for $0 < s < 1$

$$\sum_{n=0}^{\infty} P(Y_n = 0)s^n = \frac{1}{1 - \sum_{n=0}^{\infty} {}_0p_{00}^{(n)} s^n}$$

$$= \frac{1}{1 - s(1 - r(1 - g_0))(1 - s) \sum_{n=0}^{\infty} {}_0p_{00}^{(n)} s^n}.$$

Applying the Erdös–Feller–Pollard renewal theorem (see [FEL 66, p. 291]) and referring to [2.8] we have

$$\lim_{n\to\infty} P(Y_n = 0) = \frac{p - ra}{p} > 0. \qquad [2.21]$$

That $\{Y_n\}_{n\geq 0}$ is irreducible and aperiodic follows by argument similar to the one used in the proof of theorem 2.1 Therefore, taking into account [2.21], we have (see [FEL 66, p. 356]) the existence of a unique stationary distribution $\{\pi_k\}_{k\geq 0}$ such that

$$\lim_{n\to\infty} \Phi_n(s) = \Pi(s) = \sum_{k=0}^{\infty} \pi_k s^k, \qquad 0 \leq s \leq 1.$$

Letting $n \to \infty$ in equation [2.3] for the p.g.f. $\Phi_n(s)$ and referring to [2.21], we obtain [2.19].

Next, we establish the uniqueness of the solution of [2.19] among the p.g.f.s. Iterations of [2.19] yield

$$\Pi(s) = d_n(s)\Pi(f_n(s)) + \frac{p - ra}{p}(1 - d_n(s))$$
$$- \frac{r(p - ra)}{p} \sum_{k=0}^{n-1} d_k(s)(1 - g(f_k(s))). \qquad [2.22]$$

Hence, by lemma 2.1, as $n \to \infty$

$$\frac{\Pi(s)}{d_n(s)} \sim 1 + \frac{p - ra}{p} - \frac{ra}{p} + \frac{a}{p - ra}\varphi_n(s) \sim \frac{a}{p - ra}\varphi_n(s). \qquad [2.23]$$

Let $\tilde{\Pi}(s)$ be a p.g.f. solution of [2.19], which is different from $\Pi(s)$. Then it satisfies [2.22] and [2.23]. Thus,

$$\frac{\Pi(s)}{\tilde{\Pi}(s)} = \lim_{n\to\infty} \frac{\Pi(s)}{d_n(s)} \frac{d_n(s)}{\tilde{\Pi}(s)} = 1, \qquad 0 \le s \le 1,$$

proving that $\Pi(\cdot)$ and $\tilde{\Pi}(\cdot)$ coincide.

It remains to prove [2.20]. Since $\Pi(0) = (p - ra)/p$, equation [2.22] with $s = 0$ leads to

$$\Pi(f_n(0)) = \frac{p - ra}{pd_n(0)} \left[d_n(0) + r \sum_{k=0}^{n-1} d_k(0)(1 - g(f_k(0))) \right].$$

Therefore, by [2.18],

$$1 - \Pi(f_n(0)) \sim \varphi_n(0), \qquad n \to \infty.$$

Now, recalling that $1 - f_n(0) \sim (\sigma^2 n/2)^{-1}$ as $n \to \infty$, we obtain as $n \to \infty$

$$1 - \Pi(f_n(0)) \sim \begin{cases} c_\theta(s)(1 - f_n(0)) & \text{if} \quad \theta < -1 \\ c_1(s)(1 - f_n(0)) \log(1 - f_n(0)) & \text{if} \quad \theta = -1 \\ c_\theta(s)(1 - f_n(0))^{-\theta} & \text{if} \quad -1 < \theta < 0. \end{cases} \qquad [2.24]$$

On the other hand, for each $0 \le s < 1$ there exists $k \ge 0$, such that $f_k(0) \le s \le f_{k+1}(0)$ and

$$\frac{1 - \Pi(f_{k+1}(0))}{1 - f_k(0)} \le \frac{1 - \Pi(s)}{1 - s} \le \frac{1 - \Pi(f_k(0))}{1 - f_{k+1}(0)}. \qquad [2.25]$$

The claim in [2.20] follows from [2.24] and [2.25]. The proof of the theorem is complete. ∎

2.3. Regeneration and migration

In this section, we discuss an extension of the Galton–Watson process with migration, utilizing a regenerative structure. First, we recognize that a GWM process is a regenerative process. Since the probability that at least one immigrant will join the population is constant over time, the time process spent at zero (periods of emptiness for the population) is geometrically distributed. This condition is relaxed in two ways: (i) the time spent at zero is not necessary geometrically distributed and (ii) it is allowed to have an infinite first moment.

2.3.1. *Alternating regenerative processes*

Let us introduce a model of a system with replacements. More specifically, consider a machine that breaks down and is replaced or repaired. After the replacement (or repair), the machine works as new. Let $\{T_j^W\}_{j\geq 1}$ be a sequence of times when the machine works properly prior to a breakdown, and $\{T_j^R\}_{j\geq 1}$ be a sequence of time periods when the machine does not work prior to completion of a replacement or repair. Suppose these sequences are independent of each other and the random variables in each sequence are i.i.d. copies of random variables T^W and T^R, respectively. Define $T_j := T_j^R + T_j^W$ for $j = 1, 2, \ldots$ If the first working period begins at T_1^R, then the first breakdown occurs at time $T_1 = T_1^R + T_1^W$. After the replacement (or repair), the machine starts working again at $T_1 + T_2^R$ until the end of the second working period $T_1 + T_2 = T_1 + T_2^R + T_2^W$, and so on. Call T_j the jth **repairing–working cycle**, and consider the **renewal process** $N(t)$ generated by the sequence $\{T_j\}_{j\geq 1}$, that is set $S_0 = 0$, $S_n = \sum_{j=1}^n T_j$ for $n = 1, 2, \ldots$ and define for $t \geq 0$

$$N(t) = \max\{n \geq 0 : S_n \leq t\}. \qquad [2.26]$$

$N(t)$ is the number of repairing–working cycles completed by epoch t. Since t represents time, the process $\{N(t) : t \geq 0\}$ is either a discrete or continuous time stochastic process.

Let us associate with each $T_j^W = w_j$ for every $j = 1, 2, \ldots$ a stochastic process $\{Y_j(t) : 0 \leq t \leq w_j\}$, called a **cycle process**, such that

$$Y_j(0) \geq 0, \quad Y_j(t) > 0 \text{ for } 0 < t < w_j, \quad Y_j(w_j) = 0.$$

Assume that each $Y_j(t)$ has state space $(\mathbf{R}^+, \mathcal{B}(\mathbf{R}^+))$, where $\mathbf{R}^+ = [0, \infty)$ and $\mathcal{B}(\mathbf{R}^+)$ is its Borel σ-algebra. The cycles are mutually independent and stochastically equivalent to $\{Y(t) : 0 \leq t \leq w, T^W = w\}$. Also, $Y_j(t)$ may depend on T_j^W but is independent of $\{T_i^W\}_{i\neq j}$.

Finally, consider a process $\{\sigma(t) : t \geq 0\}$ defined by

$$\sigma(t) = t - S_{N(t)} - T_{N(t)+1}^R. \qquad [2.27]$$

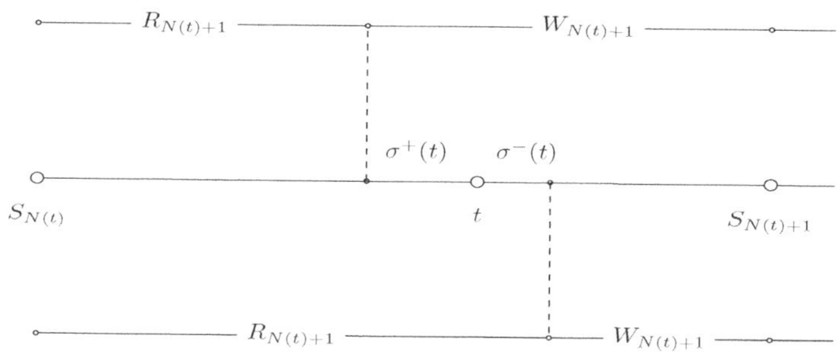

Figure 2.2. *Diagram for a renewal process*

The value of $\sigma(t)$ can be either positive or negative and $\sigma(t) = \sigma^+(t) - \sigma^-(t)$, where $\sigma^+(t) = \max\{\sigma(t), 0\}$ is the attained duration of the working time-period in progress, whereas $\sigma^-(t) = \max\{-\sigma(t), 0\}$ is the remaining time until the end of the repairing period in progress (see Figure 2.2). Now, we are in a position to define **an alternating regenerative process** $\{Z(t) : t \geq 0\}$.

DEFINITION 2.4.– *An alternating regenerative process is defined for t* ≥ 0 *by*

$$Z(t) = \begin{cases} Y_{N(t)+1}(\sigma(t)) & \textit{if } \sigma(t) \geq 0, \\ 0 & \textit{if } \sigma(t) < 0. \end{cases}$$

An example of alternating regenerative process is the regenerative process with a reward structure discussed in [WOL 89, Chapter 2]. Another example is provided by the Bellman–Harris branching processes with immigration at zero only, as studied in [MIT 02].

We need **three groups of assumptions** on the repairing period, working period and cycle process as follows:

1) For the duration of the **repairing period** T^R, we assume

$$E\left[T^R\right] < \infty \text{ or } E\left[T^R\right] = \infty \text{ and } P\left(T^R > t\right) \sim t^{-\alpha} L_1(t), \quad t \to \infty, \qquad [2.28]$$

where $\alpha \in (1/2, 1]$ and $L_1(t)$ is an s.v.f. at infinity.

2) For the duration of the **working period** T^W, we assume

$$E\left[T^W\right] = \infty \text{ and } P\left(T^W > t\right) \sim t^{-\beta} L_2(t), \quad t \to \infty, \qquad [2.29]$$

where $\beta \in (1/2, 1]$ and $L_2(t)$ is an s.v.f. at infinity.

3) For the **cycle process** $\{Y(t) : 0 \leq t \leq w;\ T^W = w\}$, we assume for $x \geq 0$

$$\lim_{t\to\infty} P\left(\frac{Y(t)}{t^\gamma L(t)} \leq x \,\middle|\, T^W > t\right) = P(Y \leq x) =: D(x), \qquad [2.30]$$

where $\gamma \geq 0$ and $L(t)$ is an s.v.f. at infinity.

The following theorem, stated without proof, is instrumental for the proof of the main result in this section.

THEOREM 2.8.– *Let $\{Z(t)\}_{t\geq 0}$ be an alternating regenerative process. Assume the conditions [2.28]–[2.30] are satisfied.*

1) If $\Delta := \lim_{t\to\infty} \dfrac{P\left(T^R > t\right)}{P\left(T^W > t\right)} < \infty$, then for $x \geq 0$,

$$\lim_{t\to\infty} P\left(\frac{Z(t)}{t^\gamma L(t)} \leq x\right) = \frac{1}{\Delta + 1}\left[\Delta + \frac{\int_0^1 D(xu^{-\gamma})u^{-\beta}(1-u)^{\beta-1}\,du}{B(1-\beta,\beta)}\right], \qquad [2.31]$$

where $B(\cdot,\cdot)$ is Euler's Beta function.

2) If $\lim_{t\to\infty} \dfrac{P(T^R > t)}{P(T^W > t)} = \infty$, then for $x \geq 0$

$$\lim_{t\to\infty} P\left(\frac{Z(t)}{t^\gamma L(t)} \leq x \,\middle|\, Z(t) > 0\right) = \frac{\int_0^1 D(xu^{-\gamma})u^{-\beta}(1-u)^{\alpha-1}\,du}{B(1-\beta,\alpha)}. \qquad [2.32]$$

2.3.2. *An extension of Galton–Watson processes with migration*

Since the state zero is a reflective barrier, the GWM process $\{Y_n\}_{n\geq 0}$ defined by [2.1] is a regenerative process. Indeed, using the notation in section 2.3.1, Y_0 stays at zero for a random period T_1^R, which is geometrically distributed with

$$P\left(T_1^R = k\right) = \left(P\left(M_1^0 = 0\right)\right)^{k-1}\left(1 - P\left(M_1^0 = 0\right)\right), \quad k = 1, 2, \ldots \qquad [2.33]$$

where $M_1^0 := M_1\mathcal{I}\{Y_0 = 0\}$ is the migration at zero. Leaving zero the process jumps up to a random level $I_{T_1^R}$ and evolves according to the rules in the model [2.1] until it hits zero again in the end of its life-period T_1^W, say. Thus, $T_1 = T_1^R + T_1^W$ forms the first regeneration period and the evolution of the process repeats, i.e. $\{Y_{n+T_1} : n \geq 0\}$ is stochastically equivalent to $\{Y_n : n \geq 0\}$.

To supply the regenerative process $\{Y_n\}_{n\geq 0}$ with a cycle process, define a migration process with no migration at zero.

DEFINITION 2.5.– *A Galton–Watson process with migration stopped at zero (GWMZ) is defined by* $Y_0^0 > 0$ *and*

$$Y_{n+1}^0 = \sum_{i=1}^{Y_n^0} X_{n,i} + M_{n+1}\mathcal{I}\{Y_n^0 > 0\}, \qquad n = 0, 1, ..., \tag{2.34}$$

(with $\sum_1^0 = 0$*), where* Y_0^0 *is independent of* $\{X_{n,i} : n = 0, 1, \dots; i = 1, 2, \dots\}$ *and* $\{M_n\}_{n\geq 1}$.

Let us introduce the sequence $\{Y_{n,j}^0\}_{j\geq 1}$ of GWMZ processes with life-periods $T_j^W = w_j$, i.e. for $j = 1, 2, \dots$

$$Y_{0,j}^0 \geq 0, \;\; Y_{n,j}^0 > 0 \text{ for } 0 < n < w_j, \;\; Y_{w_j,j}^0 = 0.$$

It is not difficult to see that the GWM process [2.1] is an alternating renewal process with jth cycle process $\{Y_{n,j}^0 : n = 0, 1, \dots w_j;\; T_j^W = w_j\}$ and jth repairing period T_j^R, where T_j^R for $j = 2, 3, \dots$ are i.i.d. copies of T_1^R in [2.33].

Next, we extend the definition of the GWM process [2.1] by weakening the conditions on its stay at zero as follows: (i) first, assume that the repairing periods $\{T_j^R\}_{j\geq 1}$ are not necessarily geometrically distributed; (ii) second, assume that $E\left[T_j^R\right]$ is not necessary finite and [2.28] holds.

DEFINITION 2.6.– *A Galton–Watson alternating regenerative process with migration* $\{Z_n\}_{n\geq 0}$ *is defined by* $Z_0 = 0$ *and for* $n = 1, 2, \dots$

$$Z_n = \begin{cases} Y_{\sigma(n),N(n)+1}^0 & \text{if } \sigma(n) \geq 0, \\ 0 & \text{if } \sigma(n) < 0, \end{cases}$$

where the cycles $\{Y_{n,j}^0\}_{j\geq 1}$ *are GWMZ processes;* $N(n)$ *and* $\sigma(n)$ *are defined by* [2.26] *and* [2.27]*, respectively.*

The queuing systems are good examples for discrete time regenerative processes. Using the well-known duality between a branching process and a M/M/1 queue (see [KEN 51]), let us describe a situation where the construction in definition 2.6 applies. Consider a queuing model in which customers arrive following a Poisson process. Then the successive times T_j from the beginning of the jth idle period to the start of the next idle period form a renewal process. Each T_j is composed of an idle (inactivity) portion T_j^R and a busy portion T_j^W, when the queue is empty and not empty, respectively. Assuming that the customers, arriving during the service time of a customer, are his/her "offspring", we obtain a Galton-Watson alternating regenerative process. The immigration component accounts for a policy when certain customers (possibly coming from a second source) accumulate and will be served after completing the service time of a "generation". Alternatively, some customers (called "emigrants") may leave the system prior to being served.

We need the following result for the cycle GWMZ $\{Y_n^0\}_{n\geq 0}$ process.

THEOREM 2.9.– *Suppose $\{Y_n^0\}_{n\geq 0}$ has offspring mean $m = 1$ and $\theta \geq 0$.*

1) Then as $n \to \infty$

$$P\left(Y_n^0 > 0\right) \sim L(n)n^{-\max\{(1-\theta),0\}}, \qquad [2.35]$$

where $L(n)$ is an s.v.f. at infinity.

2) For $x \geq 0$

$$\lim_{n\to\infty} P\left(\frac{2Y_n^0}{\sigma^2 n} \leq x \,\middle|\, Y_n^0 > 0\right) = P\left(Y^0 \leq x\right), \qquad [2.36]$$

where the limiting random variable Y^0 has a unit Exponential distribution when $0 \leq \theta \leq 1$ and a Gamma distribution with parameters θ and 1 when $\theta > 1$.

COMMENT.– If $0 \leq \theta \leq 1$, then the limit in [2.36] is unit Exponential as in the critical ordinary GW process without migration.

Recall that

$$\Delta = \lim_{n\to\infty} \frac{P\left(T^R > n\right)}{P\left(T^W > n\right)} \leq \infty.$$

Assuming that at least one of $E\left[T^R\right]$ or $E\left[T^W\right]$ is infinite, the parameter Δ can be interpreted as follows. If $\Delta < \infty$, then the life-period of the cycle process $\{Y_n^0\}_{n\geq 0}$ is asymptotically longer than the stay at zero. If $\Delta = \infty$, then, in the long run, the regenerative process with migration spends more time in state 0 than time outside state 0.

Now, we are in a position to state and prove a limit theorem for the Galton–Watson alternating regenerative processes with migration.

THEOREM 2.10.– *Let $\{Z_n\}_{n\geq 0}$ be a Galton–Watson alternating regenerative process with migration with offspring mean $m = 1$. Let* [2.28] *hold and $0 < \theta < 1/2$.*

1) If $\Delta < \infty$, then for $x \geq 0$

$$\lim_{n\to\infty} P\left(\frac{2Z_n}{\sigma^2 n} \leq x\right) = \frac{1}{\Delta+1}\left[\Delta + \frac{\int_0^1 \left(1-e^{-x/u}\right) u^{\theta-1}(1-u)^{-\theta}\,du}{B(\theta, 1-\theta)}\right].$$

The expected value of the limiting random variable is $\theta/(\Delta+1)$.

2) If $\Delta = \infty$, then for $x \geq 0$

$$\lim_{n\to\infty} P\left(\frac{2Z_n}{\sigma^2 n} \leq x \,\Big|\, Z_n > 0\right) = \frac{\int_0^1 \left(1-e^{-x/u}\right) u^{\theta-1}(1-u)^{\alpha-1}\,du}{B(\theta,\alpha)}.$$

The expected value of the limiting random variable is $\theta/(\theta+\alpha)$.

COMMENT.– The distribution of the limiting random variable is a mixture of Beta and Exponential distributions.

Proof. We will apply to $\{Z_n\}_{n\geq 0}$ theorem 2.8 for the alternating regenerative process $\{Z(t), t\geq 0\}$. To verify the assumptions of theorem 2.8, first note that [2.35] implies

$$P\left(T^W > n\right) \sim L(n)n^{-(1-\theta)}, \qquad n\to\infty,$$

and hence [2.29] holds with $\beta = 1-\theta$. Since $0 < \theta < 1/2$, by [2.36], we have

$$\lim_{n\to\infty} P\left(\frac{2Y_n^0}{\sigma^2 n} \leq x \Big| Y_n^0 > 0\right) = 1 - e^{-x}. \qquad [2.37]$$

Thus, [2.30] holds with $\gamma = 1$, $L(t) = \sigma^2/2$ and $D(x) = 1 - e^{-x}$, and so [2.31] leads to claim 1). Using the formula for the c.d.f. of the limit, it is not difficult to calculate its expected value. The proof of claim 2) is similar, taking into account [2.32] and [2.37]. ∎

It is interesting to compare theorem 2.3 for $\{Y_n\}_{n\geq 0}$ with theorem 2.10 for $\{Z_n\}_{n\geq 0}$. Recall that the former assumes a geometrically distributed T^R with $E\left[T^R\right] < \infty$, whereas in the latter T^R is not constrained to one specific distribution and might have a finite or infinite mean. If $E\left[T^R\right] = \infty$ and $\Delta = \infty$, i.e. asymptotically the repairing period is longer than the working period, the conditional on non-extinction limit in theorem 2.10, part 2), has a Beta mixture of Exponential distributions. If $E\left[T^R\right] = \infty$ but $0 \leq \Delta < \infty$, i.e. asymptotically the repairing period is shorter than the working one, the quasi-stationary distribution in theorem 2.10, part 1), is a Beta mixture of Exponential distributions with an atom at zero. It is worth mentioning the generality we gain in theorem 2.10 due to the new regeneration technique of proof in contrast to the traditional method based on p.g.f.s, used in theorem 2.3. Furthermore, the regeneration technique can be used to obtain limit theorems for $\theta \leq 0$ as well (see [YAN 03]). One limitation of the regeneration technique is that it does not apply for $\theta \geq 1/2$. Extending theorem 2.10 for $\theta \geq 1/2$ is an open problem.

2.4. Background and bibliographical notes

In the context of queuing theory, stochastic models with migration are discussed in [KEL 79]. A model of branching process with emigration–immigration (migration) is introduced in [NAG 80]. The readers are referred to the survey [VAT 93] and the monograph [RAH 95] for a detailed account of results for processes with a variety of immigration and emigration regimes. See [PAK 06] and the references within for alternative approaches to emigration.

In this chapter, we discussed the GWM processes introduced in [YAN 80]. The classification of the Markov chain $\{Y_n\}_{n\geq 0}$ is provided in [YAN 83]. Theorem 2.1 is proved in [YAN 84c]. An open question is to obtain necessary conditions for the existence of the limiting stationary distribution, like in the corresponding result for GWI from Chapter 1. Theorem 2.2 is taken from [DYA 96]. Critical GWM processes are studied (see theorem 2.3) in [YAN 80] and [YAN 83]. The supercritical GWM processes are studied in [KHA 80]. See also [GRE 88] for processes with emigration.

The model of GWM processes introduced in [YAN 80] was generalized in [YAN 96] and [YAN 00], allowing two types of emigration (family and individual) as well as state-dependent immigration.

Processes with time-dependent migration have been introduced and investigated in detail (see theorems 2.4–2.6) in [YAN 82, YAN 84a, YAN 84b, YAN 85a, YAN 85b] and [DRM 06].

The GWM processes with emigration dominating over immigration on average, i.e. $\theta < 0$, are studied in [YAN 95, YAN 96, YAN 97] and [YAN 04]. Theorem 2.7 is taken from [YAN 04]. The proof makes use of a construction for processes with pure emigration from [VAT 77]. It is interesting to note the application of these results to random walks with cookies in [KOS 08].

The introduction of the replacement model in section 2.3 follows [WOL 89]. Theorem 2.8 for the alternating regenerative process is taken from [MIT 01]. Theorem 2.9 for the GWM process with stopped at zero migration is proved in [YAN 04]. Finally, the limit result for the GWM alternating regenerative process (theorem 2.10) is from [YAN 03].

3

CB Processes: Extinction

"At each given moment there is only one layer between the 'trivial' and the impossible. Mathematical discoveries are made in this layer".

– Kolmogorov

This is the first chapter devoted to controlled branching (CB) processes treated in their entire generality. The novelty with respect to GW processes is the control over the population size by decreasing or increasing the number of individuals in each generation that participate in the future evolution of the population. The chapter starts with definition and basic properties of CB processes and proceeds with criticality classification and extinction probability results.

3.1. Definition of processes and basic properties

CB processes are discrete time stochastic processes appropriate in describing the growth of populations in which the number of reproductive individuals in each generation is subject of a control mechanism. As in the previously introduced branching models, every individual reproduces independently of the others following the same probability law called "offspring distribution". Supplementing the branching properties by controlling the number of progenitors in every generation allows for modeling a great variety of random migratory movements in and out of the population, e.g. immigration, emigration or both. Applications can be found, for example, in ecological context as follows. Consider an invasive animal species that is widely recognized as a threat to a native ecosystem, but there is disagreement about how to eradicate it. The presence of the species is appreciated by a part of the society; however, if its reproduction is left uncontrolled it will prove to be very harmful to the native ecosystem. In this case, it might be better to introduce a control mechanism over the population size in order to keep it between admissible boundaries (a deterministic control function could be implemented). Another situation prone to CB processes modeling would be the evolution of an animal population threatened by predators. In each generation, the survival of one animal (and therefore the possibility of having offspring) will be strongly affected by this threat, making the introduction of a random (e.g. binomial) control mechanism appropriate. In this situation, a Poisson approximation might be reasonable if the survival probability of an animal is low and its reproductive capacity is high (see below for details).

We proceed with the definition of CB processes. Let $(\Omega, \mathcal{A}, P)$ be a probability space on which an array of non-negative integer-valued random variables $\{X_{n,i} : n = 0, 1, \ldots; i = 1, 2, \ldots\}$ is given, where $\{X_{n,i}\}$ are i.i.d. with common p.m.f. $\{p_k\}_{k\geq 0}$. Consider a sequence of non-negative integer-valued random variables on the same probability space $\{\phi_n(k) : n, k = 0, 1, \ldots\}$, independent from $\{X_{n,i}\}_{n\geq 0, i\geq 1}$. Moreover, assume the stochastic processes $\{\phi_n(k)\}_{k\geq 0}$ for each $n \geq 0$ have the same one-dimensional probability distribution $P(\phi_0(k) = j)$ for $k, j = 0, 1, \ldots$ (time homogeneous random control).

DEFINITION 3.1.– *The CB process $\{Z_n\}_{n\geq 0}$ is defined recursively as*

$$Z_0 = N > 0, \quad Z_{n+1} = \sum_{j=1}^{\phi_n(Z_n)} X_{n,j}, \quad n \geq 0, \qquad [3.1]$$

where $\sum_1^0 = 0$ and N is an integer.

Intuitively, in the context of population dynamics, Z_n equals the number of individuals in generation n and $X_{n,j}$ is the number of offspring of the jth individual from the nth generation. Given the control mechanism $\phi_n(\cdot)$, if $\phi_n(Z_n) < Z_n$, then $Z_n - \phi_n(Z_n)$ individuals are removed from the population (this models emigration, culling process in the ecological example above, or effect of predators), and therefore do not participate in the future evolution of the process. If $\phi_n(Z_n) > Z_n$, then $\phi_n(Z_n) - Z_n$ new individuals join the population and participate in the reproduction under the same conditions as the others (this models immigration or repopulation). No control is applied to the population if $\phi_n(Z_n) = Z_n$ (see Figure 3.1).

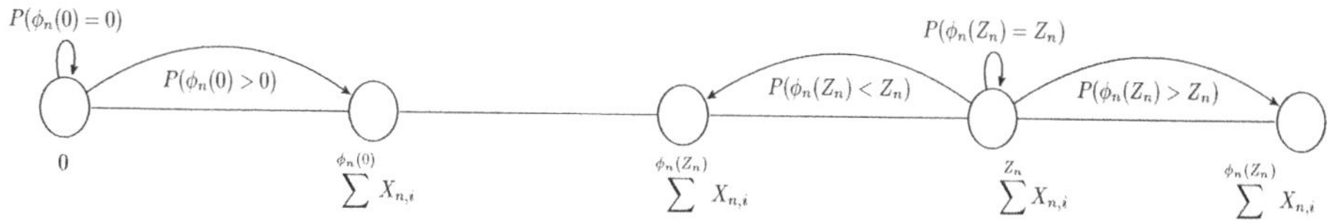

Figure 3.1. *Diagram for a CB process*

It is important to point out that a CB process does not satisfy the branching additive property, a fundamental feature of the classical GW process. Recall that, according to this property, a GW process initiated with i ancestors is equivalent to the sum of i independent copies of a GW process with single ancestor. As a result of lacking this property, contrary to the case of a GW process, the initial number of ancestors plays a significant role in the evolution of a CB process, as will be discussed later.

One way the model [3.1] can be generalized is by introducing an (finite or infinite) index set I, say, such that for each $i \in I$:

(i) $\{X_{n,j}^{(i)}\}_{n\geq 0, j\geq 1}$ are i.i.d. non-negative integer-valued random variables (independent for different values of i);

(ii) $\{\phi_n^{(i)}(k)\}_{n\geq 0, k\geq 0}$ are non-negative integer-valued random variables (independent from $\{X_{n,j}(i)\}_{n\geq 0, j\geq 1}$ for different values of i).

DEFINITION 3.2.– *The multiple controlled branching (MCB) process $\{Z_n\}_{n\geq 0}$ is defined recursively as*

$$Z_0 = N > 0, \quad Z_{n+1} = \sum_{i\in I} \sum_{j=1}^{\phi_n^{(i)}(Z_n)} X_{n,j}^{(i)}, \quad n \geq 0, \qquad [3.2]$$

where $\sum_1^0 = 0$ and N is an integer.

Particular subclasses of CB processes, some of which have been discussed in the previous chapters, can be recovered from [3.1] and [3.2] by introducing specific control variables as follows:

1) the **GW process** is obtained if $\phi_n(k) \equiv k$ for each $n, k \geq 0$;

2) the evolution of **invasive animal species** can be described by a CB process with

$$\phi_n(k) = \ell_{min}\mathcal{I}(0, \ell_{min})(k) + k\mathcal{I}[\ell_{min}, \ell_{max}](k) + \ell_{max}\mathcal{I}(\ell_{max}, \infty)(k),$$

where recall $\mathcal{I}A$ denotes the indicator function of the set A, and ℓ_{min} and ℓ_{max} are non-negative numbers, such that $\ell_{min} < \ell_{max}$;

3) the evolution of a **population threatened by predators** can be modeled by $\phi_n(k)$ distributed either binomially with parameters k and γ, for each $n, k \geq 0$, where γ is the survival probability of an individual, or having Poisson distribution with parameter γk when γ is small and the offspring mean is large;

4) the **GWI processes** are a subclass of CB processes with $\phi_n(k) = k + I_n$, for each $n, k \geq 0$, where $\{I_n\}_{n\geq 0}$ are non-negative integer-valued i.i.d. random variables, which are also independent from $\{X_{n,j}\}_{n\geq 0, j\geq 1}$. Alternatively, the GWI process can be recovered from the MCB process by setting $I = \{1, 2\}$, $\phi_n^{(1)}(k) = k$ and $\phi_n^{(2)}(k) = 1$. These two representations differ in the time point when the immigration occurs: before or after the reproduction, respectively;

5) the **FP process**, defined in Chapter 1, can be obtained from a CB process by implementing the control functions $\phi_n(k) = \max\{1, k\}$ for each $n, k \geq 0$;

6) the **GWM process**, defined in Chapter 2, is a CB process with $\phi_n(k) = \max\{0, k + M_n\}$, where $\{M_n\}_{n\geq 0}$ is a sequence of i.i.d. random variables, such that

$$P(M_n = -1) = p, \quad P(M_n = 0) = q + rP(I_n = 0),$$

and

$$P(M_n = \eta) = rP(I_n = \eta), \ \eta = 1, 2, \ldots,$$

where $p + q + r = 1$ for $p, q, r \in (0, 1)$ and $\{I_n\}_{n\geq 0}$ is the immigration component, that is, non-negative integer-valued i.i.d. random variables that are independent of $X_{n,j}$, $n = 0, 1, \ldots; j = 1, 2, \ldots$

Let us point out an important class of branching processes with a different kind of control mechanism. The so-called **population-size-dependent**

processes admit state-dependent offspring distributions and can be defined as follows:

$$Z_0 = N > 0, \quad Z_{n+1} = \sum_{j=1}^{Z_n} X_{n,j}(Z_n), \quad n \geq 0,$$

where the offspring distributions of individuals within one generation remain independently distributed but now they depend on the generation size.

Yet another peer of CB processes are the so-called **two-sex processes**. These are two-type population processes whose successive generations, $n = 0, 1, \ldots$, comprise of F_n females and M_n males, which form $Z_n = L(F_n, M_n)$ mating units. Each mating unit reproduces independently of all other mating units according to the same bivariate offspring distribution for each generation. As a result for a doubly indexed family $\{(X_{n,i}, Y_{n,i})\}_{n \geq 0, i \geq 1}$ of non-negative integer-valued i.i.d. bivariate random variables, we have

$$(F_{n+1}, M_{n+1}) = \sum_{i=1}^{Z_n} (X_{n,i}, Y_{n,i}),$$

with empty sum defined to be $(0, 0)$. The mating function $L(\cdot, \cdot)$ "controls" the number of units, which will take part in the evolution of the population.

In what follows, we focus our attention on $\{Z_n\}_{n \geq 0}$ defined by [3.1].

3.1.1. *Basic properties*

It can be proven that $\{Z_n\}_{n \geq 0}$ is a homogeneous Markov chain with stationary transition probabilities. We will establish some properties of the transition probabilities and classification of the states. To avoid trivialities we will assume that $p_0 < 1$. Depending on the distribution of the control variable $\phi_n(0)$, we will study two possibilities:

i) $P(\phi_n(0) = 0) = 1$, i.e. 0 is an absorbing state (no immigration is allowed after empty generation);

ii) $P(\phi_n(0) > 0) > 0$, i.e. 0 is not an absorbing state (there is a positive probability for immigration after empty generation).

First, let $P(\phi_n(0) = 0) = 1$ and therefore **0 is an absorbing state**. The inverse is also true as is proven in proposition 3.1.

PROPOSITION 3.1.– *Let $\{Z_n\}_{n\geq 0}$ be a CB process. The state 0 is absorbing if and only if $\phi_n(0) = 0$ a.s.*

Proof. If $P(\phi_0(0) = 0) = 1$, then by definition $P(Z_{n+r} = 0 \mid Z_n = 0) = 1$ for any $r > 0$, i.e. 0 is an absorbing state. To prove the inverse, it is sufficient to show that if $P(\phi_0(0) = 0) < 1$, then there exists $t > 0$, such that $P(\phi_n(0) = t) > 0$, i.e. 0 is not an absorbing state. Indeed, since $p_0 < 1$, there exists $k \geq 1$ such that $p_k > 0$ and hence

$$\begin{aligned} P(Z_{n+1} = kt \mid Z_n = 0) &\geq P\left(\sum_{j=1}^{t} X_{n,j} = kt\right) P(\phi_n(0) = t) \\ &\geq p_k^t P(\phi_n(0) = t) > 0. \end{aligned}$$

The proof is complete. ∎

For simplicity, we summarize the following assumptions in one.

> *(A1)* Let $P(\phi_n(0) = 0) = 1$ and at least one of the conditions holds: (i) $p_0 > 0$ (and to avoid trivialities, $p_0 < 1$); (ii) $P(\phi_0(k) = 0) > 0$ for $k = 1, 2, \ldots$

PROPOSITION 3.2.– *Let $\{Z_n\}_{n\geq 0}$ be a CB process and (A1) holds. Then the states $k = 1, 2, \ldots$ are transient and $P(Z_n \to 0) + P(Z_n \to \infty) = 1$.*

Proposition 3.2 implies that there are only two possible scenarios for a CB process verifying (A1) as $n \to \infty$: it either goes to 0 (dies out) or has unlimited growth. The proof can be seen in [YAN 75]. This is the so-called **extinction–expansion dichotomy**, which has been observed in the ordinary GW process as well.

We turn to the case when **0 is not an absorbing state**, i.e. $P(\phi_n(0) > 0) > 0$, and make two more assumptions:

> *(A2)*: $p_0 > 0$ and $p_0 + p_1 < 1$.
>
> *(A3)*: $P(\phi_n(k) > k) > 0, k = 1, 2, \ldots$

Denote the state space of the Markov chain $\{Z_n\}_{n\geq 0}$ by

$$S_N := \{i \in \mathbf{N}_0 : P(Z_m = i) > 0 \text{ for some } m \in \mathbf{N}_0;\ Z_0 = N\},$$

where $\mathbf{N}_0$ denotes the set of non-negative integers. These are the so-called "visitable states". For simplicity, we will refer to S_N merely by S. For any

$i, j \in S$, define $p_{ij}^{(n)} := P(Z_{m+n} = j \mid Z_m = i)$ for $n = 1, 2, \ldots$ For simplicity we will write p_{ij} for $p_{ij}^{(1)}$. We say that $i \in S$ **leads to** $j \in S$ if there exists a positive n such that $p_{ij}^{(n)} > 0$. The states i and j **communicate** if i leads to j and j leads to i. For $i \in S$, denote by C_i the set of all states to which i leads, i.e. $C_i := \{j \in S : p_{ij}^{(n)} > 0 \text{ for some } n \geq 1\}$.

PROPOSITION 3.3.– *Let $\{Z_n\}_{n \geq 0}$ be a CB process with $P(\phi_n(0) > 0) > 0$. If Assumptions (A2) and (A3) hold, then*

1) given a non-negative integer ν, there exists $j > \nu$ such that $j \in C_0$;

2) $C_i = C_0$ for all $i \in S$;

3) $\{Z_n\}_{n \geq 0}$ is an irreducible and aperiodic Markov chain.

Proof. 1) Assumption (A2) implies that there exists $k > 1$ such that $p_k > 0$ and since $P(\phi_n(0) > 0) > 0$, there exists a $d_0 > 0$ such that $P(\phi_n(0) = d_0) > 0$. Consequently,

$$\begin{aligned} p_{0k} &= P\left(\sum_{l=1}^{\phi_n(0)} X_{n,l} = k\right) \\ &\geq p_k p_0^{d_0 - 1} P(\phi_n(0) = d_0) > 0. \end{aligned} \quad [3.3]$$

On the other hand, taking into account Assumption (A3), it is not difficult to prove by induction on t, that

$$P(Z_{n+t} = k^{t+1} \mid Z_n = k) > 0, \quad t = 1, 2, \ldots \quad [3.4]$$

Given a non negative integer ν, there exists $t \geq 1$ such that $k^t > \nu$, and from [3.3] and [3.4], we obtain $j = k^t \in C_0$.

2) First, we will prove that $p_{i0} > 0$ for all $i \in S$, i.e. $0 \in C_i$. Indeed, if $P(Z_n = i) > 0$ for some n, then

$$\begin{aligned} p_{i0} &= P\left(\sum_{j=1}^{\phi_n(i)} X_{n,j} = 0\right) \\ &= \sum_{k=0}^{\infty} P\left(\sum_{j=1}^{k} X_{n,j} = 0\right) P(\phi_n(i) = k) \end{aligned}$$

$$= \sum_{k=0}^{\infty} p_0^k P(\phi_n(i) = k)$$

$$= E\left[p_0^{\phi_n(i)}\right] > 0.$$

If $j \in C_0$, then there exists a positive integer n such that $p_{0j}^{(n)} > 0$ and given $i \in S$ we obtain $p_{ij}^{(n+1)} \geq p_{i0}p_{0j}^{(n)} > 0$. Therefore, $j \in C_i$ and hence $C_0 \subseteq C_i$. Inversely, if $j \in C_i$, then there exists a positive integer t such that $p_{ij}^{(t)} > 0$ and $j \neq 0$ since $0 \in C_i$ for all $i \in S$. Let us prove that $j \in C_0$. If $t = 1$, then, for certain n,

$$p_{ij} = P\left(\sum_{l=1}^{\phi_n(i)} X_{n,l} = j\right)$$

$$= \sum_{\nu=1}^{\infty} P\left(\sum_{l=1}^{\nu} X_{n,l} = j\right) P(\phi_n(i) = \nu) > 0.$$

Hence, there exists $\nu > 0$ such that $P(\phi_n(i) = \nu) > 0$ and $P\left(\sum_{l=1}^{\nu} X_{n,l} = j\right) > 0$. It follows from claim 1) that there exists $h > \nu$ such that $h \in C_0$. Now, according to Assumption (A3), there exists $d_h > h$ such that $P(\phi_n(h) = d_h) > 0$. Therefore,

$$p_{hj} = P\left(\sum_{l=1}^{\phi_n(h)} X_{n,l} = j\right)$$

$$\geq P\left(\sum_{l=1}^{\nu} X_{n,l} = j\right) p_0^{d_h - \nu} P(\phi_n(h) = d_h) > 0,$$

i.e. $j \in C_h$ and, since $h \in C_0$, we obtain $j \in C_0$. If $t > 1$ for certain n, then

$$p_{ij}^{(t)} = \sum_{h=0}^{\infty} p_{ih} p_{hj}^{(t-1)} > 0$$

and therefore there must exist $h \geq 0$, such that

$$p_{hj}^{(t-1)} > 0 \text{ and } p_{ih} > 0.$$

If $h = 0$, then $p_{0j}^{(t-1)} > 0$, i.e. $j \in C_0$. Otherwise, since $p_{ih} > 0$, using a similar argument as that for $t = 1$, we obtain $h \in C_0$. Therefore, since $p_{hj}^{(t-1)} > 0$, we have $j \in C_0$.

3) It follows from claim 2) that if one state leads to another state, then 0 leads to it too. Therefore, the state space consists of only one essential class, the class of 0. In other words, any state in C_0 communicates with those states that it leads to and any non-null state leads to 0 in one step. Let us prove that the states are aperiodic. Indeed, since $p_0 + p_1 < 1$, there exists $k > 1$ such that $p_k > 0$. Assumption (A3) yields the existence of $d_k > k$ such that $P(\phi_n(k) = d_k) > 0$ and by Assumption (A2) we have $p_0 > 0$. Then,

$$
\begin{aligned}
p_{kk} &\geq P\left(\sum_{\nu=1}^{d_k} X_{n,\nu} = k\right) P(\phi_n(k) = d_k) \\
&\geq p_k p_0^{d_k - 1} P(\phi_n(k) = d_k) > 0.
\end{aligned}
$$

Hence, $\{Z_n\}_{n \geq 0}$ is an irreducible and aperiodic Markov chain. ∎

3.1.2. *Probability generating functions and moments*

Investigations based on probability generating functions (**p.g.f.s**) play a very important role in the theory of discrete-time branching processes including GW and GWI processes, as was pointed out in Chapter 1. Unfortunately, the p.g.f.s are of not so much help in the analysis of the general CB processes. Nevertheless, it is worth establishing relationships between p.g.f.s of certain random variables associated with a CB process and applying them to calculate some moments of Z_n. Assume $s \in [0, 1]$. Let $f(s) = \sum_{k=0}^{\infty} p_k s^k$ be the **offspring p.g.f.**; $g_k(s) := \sum_{l=0}^{\infty} P(\phi_0(k) = l) s^l$ be the **control p.g.f.**; $F_n(s) = \sum_{k=0}^{\infty} P(Z_n = k) s^k$ be the **process p.g.f.**; and for each $k \geq 0$, $h_k(s) := \sum_{j=0}^{\infty} P(Z_{n+1} = j | Z_n = k) s^j$ denote the **transition p.g.f.s**. The following proposition connects transition, offspring and control generating functions.

PROPOSITION 3.4.– *Let $\{Z_n\}_{n \geq 0}$ be a CB process. Then for $0 \leq s \leq 1$*

$$
h_k(s) = g_k(f(s)), \quad k = 0, 1, \ldots
$$

Proof. By definition, for $k = 0, 1, \ldots$

$$
\begin{aligned}
h_k(s) &= E\left[s^{\sum_{j=1}^{\phi_n(k)} X_{n,j}}\right] \\
&= E\left[E\left[s^{\sum_{j=1}^{\phi_n(k)} X_{n,j}} \mid \phi_n(k)\right]\right] \\
&= \sum_{l=0}^{\infty} E\left[s^{\sum_{j=1}^{l} X_{n,j}}\right] P(\phi_n(k) = l) \\
&= \sum_{l=0}^{\infty} \left(E\left[s^{X_{0,1}}\right]\right)^l P(\phi_n(k) = l) \\
&= E\left[f(s)^{\phi_n(k)}\right] \\
&= g_k(f(s)).
\end{aligned}
$$

■

Note that, $h_k(s)$ for every k is the p.g.f. of the random variable $\sum_{j=1}^{\phi_n(k)} X_{n,j}$. Consequently, if $\phi_n(0) = 0$ a.s., then $h_0(s) = 1$ for $0 \leq s \leq 1$.

As in the previous chapters, let m and σ^2 denote the offspring mean and variance, respectively. Unless stated otherwise, we assume σ^2 (and thus m) to be finite. Denote

$$\varepsilon(k) := E[\phi_n(k)] \quad \text{and} \quad \nu^2(k) := Var[\phi_n(k)], \quad k = 0, 1, \ldots$$

Also define $\mathcal{F}_n := \sigma(Z_0, \ldots, Z_n)$ to be the σ-algebra generated by the random variables $Z_0, \ldots, Z_n$, for $n = 0, 1, \ldots$.

We will calculate the mean and the variance of the process $\{Z_n\}_{n\geq 0}$.

PROPOSITION 3.5.– *Let $\{Z_n\}_{n\geq 0}$ be a CB process. Then*

1) $E[Z_{n+1} \mid \mathcal{F}_n] = m\varepsilon(Z_n)$ *a.s.;*

2) $Var[Z_{n+1} \mid \mathcal{F}_n] = m^2\nu^2(Z_n) + \sigma^2\varepsilon(Z_n)$ *a.s..*

Proof. 1) Using proposition 3.4, we obtain for $k = 0, 1, \ldots$

$$
\begin{aligned}
E[Z_{n+1} \mid Z_n = k] &= h'_k(1) \\
&= g'_k(1) f'(1) \\
&= \varepsilon(k) m.
\end{aligned}
$$

2) Applying proposition 3.4 again, we have for $k = 0, 1, \ldots$

$$\begin{aligned} Var[Z_{n+1} \mid Z_n = k] &= Var\left[\sum_{j=1}^{\phi_n(k)} X_{n,j}\right] \\ &= h_k''(1) + h_k'(1) - (h_k'(1))^2 \\ &= Var[X_{01}]E[\phi_n(k)] + E[X_{0,1}]^2 Var[\phi_n(k)] \\ &= \sigma^2 \varepsilon(k) + m^2 \nu^2(k). \end{aligned}$$

■

COMMENT.– Proposition 3.5 yields

$$\begin{aligned} E[Z_{n+1}] &= mE[\varepsilon(Z_n)], \\ Var[Z_{n+1}] &= E[Var[Z_{n+1} \mid \mathcal{F}_n]] + Var[E[Z_{n+1} \mid \mathcal{F}_n]] \\ &= \sigma^2 E[\varepsilon(Z_n)] + m^2(E[\nu^2(Z_n)] + Var[\varepsilon(Z_n)]). \end{aligned}$$

3.2. Extinction probability

In this section, we discuss the extinction problem for CB processes and find conditions for almost sure extinction as well as for positive non-extinction probability. Assume **state 0 is absorbing** or equivalently $P(\phi_n(0) = 0) = 1$. Moreover, to avoid trivialities, assume condition (A1) holds. According to proposition 3.2, we have

$$P(Z_n \to 0 \mid Z_0 = N) + P(Z_n \to \infty \mid Z_0 = N) = 1, \quad N > 1. \quad [3.5]$$

We will see later that the initial number of individuals plays a significant role in the evolution of the process. Bearing this in mind, denote

$$\begin{aligned} q_N &:= P(\text{extinction} \mid Z_0 = N) \\ &= P(Z_n \to 0 \mid Z_0 = N). \end{aligned}$$

Let us also introduce the **mean growth rates** $\tau_m(k)$ for $k = 1, 2, \ldots$ by

$$\begin{aligned}\tau_m(k) &:= \frac{1}{k} E[Z_{n+1}|Z_n = k] \\ &= \frac{\varepsilon(k)}{k} m.\end{aligned}$$

The mean growth rate for a particular k is intuitively interpreted as the expected growth rate per individual when, in a certain generation, there are k individuals. Note that for a GW process, $\tau_m(k) = m$ for every k. The sequence $\{\tau_m(k)\}_{k\geq 1}$ plays a key role for the evolution of the process. Depending on the asymptotic behavior of $\tau_m(k)$, three classes of CB processes can be identified.

DEFINITION 3.3.– *A CB process is said to be*

(i) subcritical if $\limsup_{k\to\infty} \tau_m(k) < 1$;

(ii) critical if $\liminf_{k\to\infty} \tau_m(k) \leq 1 \leq \limsup_{k\to\infty} \tau_m(k)$;

(iii) supercritical if $\liminf_{k\to\infty} \tau_m(k) > 1$.

The results in this chapter and the following one will reveal the similarities and differences between the CB and the ordinary GW processes.

3.2.1. *Subcritical processes*

In the subcritical case, the CB processes and the ordinary GW processes have similar asymptotic behavior. In particular, in both the extinction is certain.

THEOREM 3.1.– *If* $\{Z_n\}_{n\geq 0}$ *is a subcritical CB process, then* $q_N = 1$ *for all* $N \geq 1$.

Proof. If $\limsup_{k\to\infty} \tau_m(k) < 1$, then there exist $N_0 \geq 1$ and α satisfying $0 < \alpha < m^{-1}$, such that

$$\sup_{k\geq N_0} \frac{\varepsilon(k)}{k} \leq \alpha.$$

Let $N \geq 1$. We will prove that the sequence $\{E[Z_{n+1} \mid Z_0 = N]\}_{n \geq 0}$ is bounded. Indeed,

$$
\begin{aligned}
E[Z_{n+1} \mid Z_0 = N] &= \sum_{l=0}^{\infty} E[Z_{n+1} \mid Z_n = l] P(Z_n = l \mid Z_0 = N) \\
&= \sum_{l=0}^{N_0 - 1} E[Z_{n+1} \mid Z_n = l] P(Z_n = l \mid Z_0 = N) \\
&\quad + \sum_{l=N_0}^{\infty} E[Z_{n+1} \mid Z_n = l] P(Z_n = l \mid Z_0 = N) \\
&\leq m \sum_{l=0}^{N_0 - 1} \sup_{0 \leq t \leq N_0 - 1} \varepsilon(t) P(Z_n = l \mid Z_0 = N) \\
&\quad + m\alpha \sum_{l=N_0}^{\infty} l P(Z_n = l \mid Z_0 = N) \\
&\leq m\varepsilon_{N_0} + m\alpha E[Z_n \mid Z_0 = N],
\end{aligned}
$$

where $\varepsilon_{N_0} = \sup_{0 \leq t \leq N_0 - 1} \varepsilon(t)$. Hence, after iterating, we have for any $n \geq 1$

$$
\begin{aligned}
E[Z_{n+1} \mid Z_0 = N] &\leq m\varepsilon_{N_0} \sum_{j=0}^{n-1} (\alpha m)^j + (\alpha m)^n N \\
&\leq \frac{m\varepsilon_{N_0}}{1 - \alpha m} + N < \infty.
\end{aligned}
$$

Applying Chebyshev's inequality, we obtain for given $M > 0$

$$
\begin{aligned}
P(Z_n \geq M \mid Z_0 = N) &\leq \frac{E[Z_n \mid Z_0 = N]}{M} \\
&\leq \frac{1}{M} \left(\frac{m\varepsilon_{N_0}}{1 - \alpha m} + N \right).
\end{aligned}
$$

Thus, for given $\epsilon > 0$, there exists M_ϵ, such that for any $M \geq M_\epsilon$,

$$
P(Z_n \geq M \mid Z_0 = N) < \frac{\epsilon}{2}.
$$

Assumption (A1) yields, for any $k > 0$,

$$\lim_{n \to \infty} P(Z_n = k \mid Z_0 = N) = 0.$$

Consequently, for given $M > 1$, there exists n_ϵ, such that, for any $n \geq n_\epsilon$,

$$P(1 \leq Z_n < M \mid Z_0 = N) < \frac{\epsilon}{2}.$$

Choosing $n \geq n_\epsilon$ and $M \geq M_\epsilon$, we obtain

$$\begin{aligned} P(Z_n = 0 \mid Z_0 = N) &= 1 - P(Z_n \geq M \mid Z_0 = N) \\ &\quad - P(1 \leq Z_n < M \mid Z_0 = N) \\ &\geq 1 - \epsilon. \end{aligned}$$

Thus, $P(Z_n \to 0 \mid Z_0 = N) = 1$ for any $N > 0$. ■

3.2.2. *Supercritical processes*

In the supercritical case, the control variables have much more influence than in the subcritical case. While supercritical ordinary GW processes always have a positive non-extinction probability, this is no longer true for CB processes. As an illustration, we present the following example.

EXAMPLE 3.1.– Let $0 < \epsilon < 1$. Consider a CB process, such that $\phi_n(0) = 0$ a.s. and for each $k > 0$,

$$P(\phi_n(k) = 0) = \epsilon^{1/k}, \quad P\left(\phi_n(k) = k\left[\frac{1}{1 - \epsilon^{1/k}}\right]\right) = 1 - \epsilon^{1/k}, \quad [3.6]$$

where $[x]$ denotes the integer part of x. It is clear that the sequence of control random variables converges in distribution to 0. Figure 3.2 presents the control p.g.f.s for different values of k and $\epsilon = 0.05$. We will prove that $\lim_{k \to \infty} \varepsilon(k)/k = 1$. Indeed,

$$\begin{aligned} \frac{\varepsilon(k)}{k} &= \frac{E[\phi_n(k)]}{k} \\ &= (1 - \epsilon^{1/k}) \left[\frac{1}{1 - \epsilon^{1/k}}\right]. \end{aligned}$$

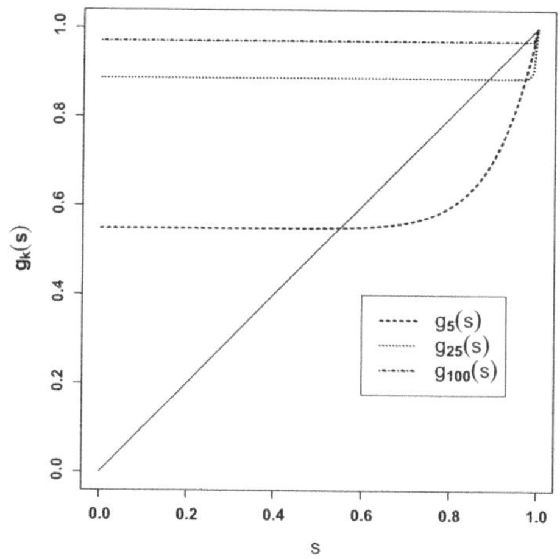

Figure 3.2. *Control probability generating functions for particular cases in* [3.6].

Set $x_k = 1 - \epsilon^{1/k}$ for $k = 1, 2 \ldots$ The sequence $\{x_k\}_{k \geq 1}$ is non-increasing, converges to 0, and $0 < x_k < 1$. For each x_k, there exists $n_k \in \mathbf{N}^+$, such that

$$\frac{1}{n_k} < x_k \leq \frac{1}{n_k - 1},$$

or equivalently

$$n_k - 1 \leq \frac{1}{x_k} < n_k.$$

Hence, $[x_k^{-1}] = n_k - 1$ and

$$1 - \frac{1}{n_k} < x_k \left[\frac{1}{x_k} \right] \leq 1.$$

Therefore, $\lim_{k \to \infty} \varepsilon(k)/k = \lim_{k \to \infty} x_k [x_k^{-1}] = 1$. Next, we will prove by induction that for $n = 1, 2, \ldots$

$$P(Z_n > 0 \mid Z_0 = N) < (1 - \epsilon)^n. \qquad [3.7]$$

Note that for $n = 0, 1, \ldots$

$$P(Z_{n+1} > 0 \mid Z_n = k) = 1 - P\left(\sum_{i=1}^{\phi_n(k)} X_{n,i} = 0 \right)$$

$$\leq 1 - P(\phi_n(k) = 0)$$
$$= 1 - \epsilon^{1/k}.$$

To verify [3.7] for $n = 1$, we observe that

$$\begin{aligned} P(Z_1 > 0 \mid Z_0 = N) &\leq 1 - \epsilon^{1/N} \\ &< 1 - \epsilon. \end{aligned}$$

Suppose [3.7] is true for $n - 1$. Then

$$\begin{aligned} P(Z_n > 0 \mid Z_0 = N) &= \sum_{j=1}^{\infty} P(Z_n > 0 \mid Z_{n-1} = j) \\ &\quad \cdot P(Z_{n-1} = j \mid Z_0 = N) \\ &\leq \sum_{j=1}^{\infty} (1 - \epsilon^{1/j}) P(Z_{n-1} = j \mid Z_0 = N) \\ &< (1 - \epsilon)(1 - \epsilon)^{n-1} \\ &= (1 - \epsilon)^n, \end{aligned}$$

which completes the proof by induction. Therefore, choosing an offspring distribution with mean $m > 1$, we have $\liminf_{k \to \infty} \tau_m(k) > 1$, i.e. the CB process is supercritical but nevertheless $q_N = 1$ for all $N \geq 1$, according to [3.7]. Figure 3.3 shows the evolution of three simulated processes, satisfying the conditions of this example with $N = 1$ and Poisson offspring distribution. On the top is the evolution of a CB process with $m = 1.005$ and $\epsilon = 5 \times 10^{-20}$; the middle graph shows the evolution of a CB process with $m = 1.01$ and $\epsilon = 5 \times 10^{-17}$; on the bottom is the evolution of a CB process with $m = 1.03$ and $\epsilon = 5 \times 10^{-10}$. The processes die out in generations 123, 90 and 50, respectively. Notice that, the larger ϵ and m are, the shorter the process' lifetime is. □

Example 3.1 shows that extinction criteria cannot be formulated in terms of $\tau_m(k)$ only. Additional conditions are needed to guarantee a positive non-extinction probability for supercritical CB processes. In theorem 3.2 below, such extra conditions are imposed in terms of the first two moments of the control functions.

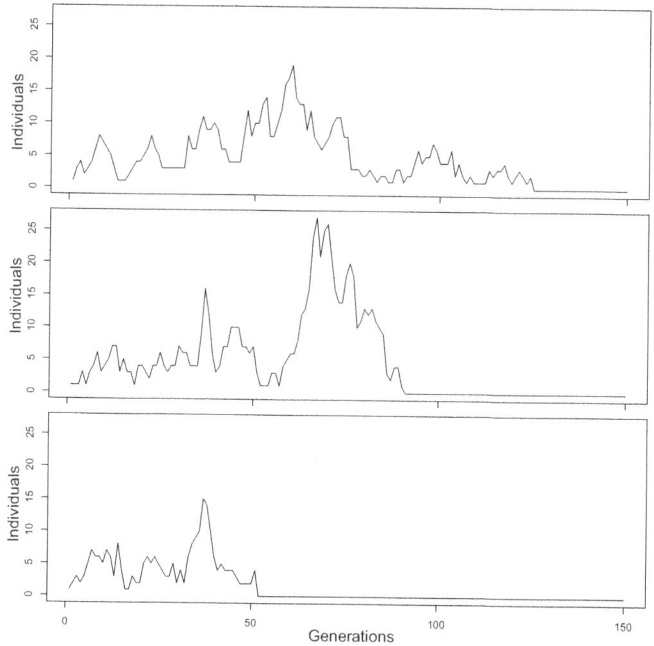

Figure 3.3. *Evolution of three simulated CB processes*

THEOREM 3.2.– *Let $\{Z_n\}_{n\geq 0}$ be a supercritical CB process. Assume the following two sequences are bounded*

$$\left\{\frac{\varepsilon(k)}{k}\right\}_{k\geq 1} \qquad \textit{and} \qquad \left\{\frac{\nu^2(k)}{k}\right\}_{k\geq 1}.$$

Then there exists $N_0 \geq 1$ such that $q_N < 1$ for all $N \geq N_0$.

Proof. By the definition of supercritical CB process, there exist $k_0 > 0$ and α with $\alpha m > 1$, such that for $k \geq k_0$

$$\frac{\varepsilon(k)}{k} \geq \alpha.$$

We will prove that there exists a positive integer N_0 such that for $N \geq N_0$

$$P(Z_n \to \infty | Z_0 = N) > 0. \qquad [3.8]$$

Construct the truncated random variables $X^a_{n,j} := \min\{X_{n,j}, a\}$ for $n = 0, 1, \ldots$ and $j = 1, 2, \ldots$, where a is a positive constant. Define $\{Z^a_n\}_{n\geq 0}$ recursively by

$$Z^a_0 = Z_0, \qquad Z^a_{n+1} = \sum_{j=1}^{\phi_n(Z_n)} X^a_{n,j} \qquad n = 0, 1, \ldots$$

It is clear that $Z^a_{n+1} \leq Z_{n+1}$ for $n = 0, 1, \ldots$. If $m_a := E[X^a_{0,1}]$, then there exist a and ϵ such that $1 < \alpha m_a < \alpha m$ and $0 < \epsilon < \alpha m_a - 1$. To prove [3.8], it is sufficient to show that for certain $N_0 \geq 1$ and for all $N \geq N_0$

$$P\left(\bigcap_{n=0}^{\infty} \{Z^a_{n+1} > (\alpha m_a - \epsilon)Z_n\} \middle| Z_0 = N\right) > 0. \qquad [3.9]$$

Denote $A_n := \{Z^a_{n+1} > (\alpha m_a - \epsilon)Z_n\}$ for $n = 0, 1, \ldots$. Observing that for $N \geq k_0$

$$\begin{aligned} E[Z^a_{n+1} \mid Z_n = N] &= \varepsilon(N) m_a \\ &\geq \alpha m_a N \end{aligned}$$

and using Markov's inequality, we obtain for $N \geq k_0$

$$\begin{aligned} P(A^c_n \mid Z_n = N) &\leq P(Z^a_{n+1} \leq \varepsilon(Z_n) m_a - \epsilon Z_n \mid Z_n = N) \qquad [3.10] \\ &\leq P(|Z^a_{n+1} - \varepsilon(Z_n) m_a| \geq \epsilon Z_n \mid Z_n = N) \\ &\leq \frac{1}{\epsilon^2 N^2} Var[Z^a_{n+1} \mid Z_n = N] \\ &= \frac{1}{\epsilon^2 N^2} \left(Var[X^a_{0,1}]\varepsilon(N) + m_a^2 \nu^2(N)\right) \\ &\leq \frac{1}{\epsilon^2 N} \left(Var[X^a_{0,1}] \sup_{k\geq 1} \frac{\varepsilon(k)}{k} + m_a^2 \sup_{k\geq 1} \frac{\nu^2(k)}{k}\right) \\ &=: \frac{C_0}{N}, \qquad \text{say.} \end{aligned}$$

Now, for $N \geq k_0$ and $k \geq 1$ we have

$$P\left(\bigcap_{n=0}^{k} A_n \middle| Z_0 = N\right)$$

$$= P(A_0 \mid Z_0 = N) \prod_{n=1}^{k} P\left(A_n \middle| \bigcap_{l=0}^{n-1} A_l \cap \{Z_0 = N\} \right).$$

From [3.10], since $Z_n^a \le Z_n$ and $\alpha m_a - \epsilon > 1$, we obtain for $n = 1, \ldots, k$

$$\begin{aligned} P\left(A_n \middle| \bigcap_{l=0}^{n-1} A_l \cap \{Z_0 = N\} \right) &\ge \inf_{t > N(\alpha m_a - \epsilon)^n} P(A_n \mid Z_n = t) \\ &\ge 1 - \frac{C_0}{N} \frac{1}{(\alpha m_a - \epsilon)^n}. \end{aligned}$$

Hence, choosing $N_0 = [\max\{k_0, C_0\}] + 1$, where recall that $[x]$ denotes the integer part of x, and taking into account $\sum_{n=0}^{\infty} (\alpha m_a - \epsilon)^{-n} < \infty$, we have for $N \ge N_0$

$$\begin{aligned} P\left(\bigcap_{n=0}^{\infty} A_n \middle| Z_0 = N \right) &= \lim_{k \to \infty} P\left(\bigcap_{n=0}^{k} A_n \middle| Z_0 = N \right) \\ &\ge \prod_{n=0}^{\infty} \left(1 - \frac{C_0}{N} \frac{1}{(\alpha m_a - \epsilon)^n} \right) > 0, \end{aligned}$$

which completes the proof. ∎

Let us verify that the assumption in theorem 3.2 on the sequence of variances fails in example 3.1. Indeed,

$$\begin{aligned} \nu^2(k) &= E[\phi_n^2(k)] - \varepsilon^2(k) \\ &= k^2 \left[\frac{1}{1 - \epsilon^{1/k}} \right]^2 (1 - \epsilon^{1/k}) - k^2 \left[\frac{1}{1 - \epsilon^{1/k}} \right]^2 (1 - \epsilon^{1/k})^2 \\ &= k^2 \left[\frac{1}{1 - \epsilon^{1/k}} \right]^2 (1 - \epsilon^{1/k}) \epsilon^{1/k}. \end{aligned}$$

Taking into account that $\lim_{k \to \infty} \epsilon^{1/k} = 1$, $\lim_{k \to \infty} (1 - \epsilon^{1/k})[(1 - \epsilon^{1/k})^{-1}] = 1$, and $\lim_{k \to \infty} k[(1 - \epsilon^{1/k})^{-1}] = \infty$, it is immediate that

$$\lim_{k \to \infty} \frac{\nu^2(k)}{k} = \infty.$$

Hence, $\{\nu^2(k)/k\}_{k \ge 1}$ is not bounded.

3.2.3. *Critical processes*

In this section, we are concerned with the critical case, that is, when

$$\liminf_{k\to\infty} \tau_m(k) \leq 1 \leq \limsup_{k\to\infty} \tau_m(k).$$

However, these asymptotic conditions on the mean growth rates cannot be used as criteria for extinction. What counts is the speed of convergence. As could be expected, if $\{\tau_m(k)\}_{k\geq 1}$ converges to 1 fast enough, then the critical CB process behaves asymptotically as an ordinary critical GW process and the extinction is almost certain. If, on the other hand, $\{\tau_m(k)\}_{k\geq 1}$ converges slow-enough, the critical CB process develops as a supercritical GW process, having a positive probability of non–extinction (although its behavior on the non–extinction set is different as we will see in the next chapter). We will make this precise in what follows.

Observe that a CB process $\{Z_n\}_{n\geq 0}$ defined by [3.1] satisfies the following stochastic difference equation

$$\begin{aligned} Z_{n+1} &= Z_n + E[Z_{n+1} - Z_n \mid Z_n] + Z_{n+1} - E[Z_{n+1} \mid Z_n] \\ &= Z_n + g(Z_n) + \xi_{n+1}, \quad n = 0, 1, \ldots \end{aligned} \qquad [3.11]$$

where for $k = 0, 1, \ldots$

$$\begin{aligned} g(k) &:= E[Z_{n+1} - Z_n | Z_n = k] \\ &= m\varepsilon(k) - k \end{aligned}$$

is the mean difference of two consecutive population sizes, and

$$\xi_{n+1} := Z_{n+1} - E[Z_{n+1} \mid Z_n]$$

is a random variable equal to the population size deviation from its mean. It is not difficult to see that $\{\xi_n\}_{n\geq 1}$ are zero-mean, square-integrable martingale differences with respect to the family of σ-algebras $\mathcal{F}_n = \sigma(Z_0, \ldots, Z_n)$, $n = 0, 1, \ldots$ (see definition A1.1 in Appendix 1). Denote for $\zeta > 0$

$$\ell_\zeta(k) := E[|\xi_{n+1}|^\zeta \mid Z_n = k], \qquad k = 1, 2, \ldots$$

In particular, the conditional variance $\ell_2(k)$ depends on the present state of the process only. In fact, as it was shown in proposition 3.5, for $k = 1, 2, \ldots$

$$\ell_2(k) = m^2\nu^2(k) + \sigma^2\varepsilon(k).$$

The following result provides sufficient conditions for almost sure extinction.

THEOREM 3.3.– *Let $\{Z_n\}_{n\geq 0}$ be a critical CB process, such that:*

(i) $\tau_m(k) = 1 + o(1)$ *as* $k \to \infty$ *and* $\tau_m(k) \geq 1$, $k = 1, 2 \ldots$;

(ii) for some $0 < \delta \leq 1$

$$\ell_{2+\delta}(k) = o(\ell_2(k)k^\delta) \quad as \quad k \to \infty;$$

(iii)
$$\limsup_{k\to\infty} \frac{2kg(k)}{\ell_2(k)} < 1.$$

Then $q_N = 1$ *for any* $N = 1, 2, \ldots$

Sufficient condition for a positive non-extinction probability is given in theorem 3.4.

THEOREM 3.4.– *Let $\{Z_n\}_{n\geq 0}$ be a critical CB process, such that:*

(i) $\tau_m(k) = 1 + o(1)$ *as* $k \to \infty$;

(ii) for some $0 < \delta \leq 1$ *and* $s > 1$

$$\ell_{2+\delta}(k) = o\left(\frac{k^{1+\delta}g(k)}{(\log k)^s}\right) \quad as \quad k \to \infty;$$

(iii)
$$\liminf_{k\to\infty} \frac{2kg(k)}{\ell_2(k)} > 1.$$

Then $q_N < 1$ *for any* $N = 1, 2, \ldots$

COMMENT.– (a) Note that, if assumption (iii) in theorem 3.3 is true, then a sufficient condition for (ii) in theorem 3.3 to hold is $\ell_{2+\delta}(k) = o(g(k)k^{1+\delta})$ because

$$\frac{\ell_{2+\delta}(k)}{\ell_2(k)k^\delta} = \frac{\ell_{2+\delta}(k)}{g(k)k^{1+\delta}} \cdot \frac{kg(k)}{\ell_2(k)}.$$

Analogously, if assumption (iii) in theorem 3.4 is true, then a sufficient condition for (ii) in theorem 3.4 to hold is $\ell_{2+\delta}(k) = o(\ell_2(k)k^\delta(\log k)^{-s})$.

Assumption (iii) in both theorems can be interpreted as a convergence rate condition due to the fact that

$$\frac{kg(k)}{\ell_2(k)} = \frac{\tau_m(k) - 1}{\ell_2(k)k^{-2}}.$$

(b) From a practical point of view, it is of interest to find simple conditions validating the hypotheses about $(2+\delta)$th absolute conditional moment of ξ_{n+1}. To this end, using

$$|a+b|^r \leq C_r(|a|^r + |b|^r) \text{ for } r > 0,$$

where C_r is a positive constant (**C_r-inequality**) and Marcinkiewicz–Zygmund's inequality (see lemma A2.1 in Appendix 2), we obtain

$$\ell_{2+\delta}(k) \leq C_{2+\delta}\left(E\left[\left|\sum_{i=1}^{\phi_n(k)} (X_{n,i} - m)\right|^{2+\delta}\right] \right.$$

$$\left. + m^{2+\delta} E\left[|\phi_n(k) - \varepsilon(k)|^{2+\delta}\right]\right)$$

and as $k \to \infty$

$$E\left[\left|\sum_{i=1}^{\phi_n(k)} (X_{n,i} - m)\right|^{2+\delta}\right] = O\left(E\left[(\phi_n(k))^{1+\delta/2}\right]\right).$$

Hence, the hypotheses on $E\left[|\xi_{n+1}|^{2+\delta} \mid Z_n = k\right]$ can be validated by regularity conditions on $E\left[(\phi_0(k))^{1+\delta/2}\right]$ and $E\left[|\phi_0(k) - \varepsilon(k)|^{2+\delta}\right]$.

A key role in the proofs of theorems 3.3 and 3.4 is played by the following result from [GON 05d].

PROPOSITION 3.6.– *Let $\{X_n\}_{n\geq 0}$ be a sequence of non-negative random variables and let $\{\mathcal{F}_n\}_{n\geq 0}$ be a non-decreasing sequence of σ-algebras, such that X_n is $\mathcal{F}_n$–measurable. Let f be a positive function on $\mathbf{R}^+$.*

1) If f is increasing and for some constant $A > 0$,

$$E[f(X_{n+1}) \mid \mathcal{F}_n] \leq f(X_n) \text{ a.s. on } \{X_n > A\},\ n = 0, 1, \ldots,$$

then $P(X_n \to \infty) = 0$.

2) Suppose that for any constant C^*, there exists a positive integer n such that $P(X_n > C^*) > 0$ and moreover $P(X_n \to 0) + P(X_n \to \infty)$=1. If f is decreasing and for some constant $A > 0$,

$$E[f(X_{n+1}) \mid \mathcal{F}_n] \leq f(X_n) \text{ a.s. on } \{X_n > A\},\ n = 0, 1, \ldots,$$

then $P(X_n \to \infty) > 0$.

Now we are in a position to prove theorems 3.3 and 3.4.

Proof of theorem 3.3. It follows from [3.11] that

$$\log(Z_{n+1} + 1) = \log(Z_n + 1) + \log\left(1 + \frac{g(Z_n)}{Z_n + 1} + \frac{\xi_{n+1}}{Z_n + 1}\right)$$

and taking expectation of sides

$$\begin{aligned} E[\log(Z_{n+1} + 1) | Z_n = k] &= \log(k + 1) \\ &\quad + E\left[\log\left(1 + \frac{g(Z_n)}{Z_n + 1} + \frac{\xi_{n+1}}{Z_n + 1}\middle| Z_n = k\right)\right]. \end{aligned} \qquad [3.12]$$

If $x > -1$ and $0 \leq \delta \leq 1$, then the following elementary inequality holds

$$\log(x + 1) \leq x - \frac{x^2}{2} + \frac{1}{2}|x|^{2+\delta}. \qquad [3.13]$$

Applying [3.13] with $x = \dfrac{g(Z_n)}{Z_n + 1} + \dfrac{\xi_{n+1}}{Z_n + 1}$ to the right-hand side of [3.12], using the C_r-inequality and $g(k) - k(\gamma_m(k) - 1) \geq 0$, we obtain, for certain positive constant $C_{2+\delta}$ and $k = 1, 2, \ldots$

$$\begin{aligned} E[\log(Z_{n+1} + 1) \,|\, Z_n = k] &\leq \log(k + 1) + \frac{g(k)}{k} \\ &\quad - \frac{g^2(k)}{2(k + 1)^2} - \frac{\ell_2(k)}{2(k + 1)^2} \\ &\quad + \frac{1}{2} C_{2+\delta} \left(\frac{g^{2+\delta}(k)}{k^{2+\delta}} + \frac{\ell_{2+\delta}(k)}{k^{2+\delta}}\right) \\ &= \log(k + 1) + \frac{g(k)}{k}\Delta_1(k) - \frac{\ell_2(k)}{2k^2}\Delta_2(k), \end{aligned}$$

where

$$\Delta_1(k) = 1 - \frac{g(k)k}{2(k+1)^2} + \frac{1}{2}C_{2+\delta}\frac{g^{1+\delta}(k)}{k^{1+\delta}}$$

and

$$\Delta_2(k) = \frac{k^2}{(k+1)^2} - \frac{1}{2}C_{2+\delta}\frac{\ell_{2+\delta}^{1+\delta}(k)}{\ell_2(k)k^\delta}.$$

It follows from (i) and (ii) that $\lim_{k\to\infty}\Delta_1(k) = \lim_{k\to\infty}\Delta_2(k) = 1$, so for a large enough k and a sufficiently small $\epsilon > 0$, we have

$$\frac{g(k)}{k}\Delta_1(k) - \frac{\ell_2(k)}{2k^2}\Delta_2(k) \le \frac{g(k)}{k}(1+\epsilon) - \frac{\ell_2(k)}{2k^2}(1-\epsilon).$$

From (iii), choosing ϵ such that $0 < \epsilon < (1-b)/(1+b)$, where

$$b := \limsup_{k\to\infty}\frac{2kg(k)}{\ell_2(k)},$$

the right-hand side in the previous inequality becomes negative. Therefore, if $Z_n \ge A$, for some large enough constant A, then

$$E[\log(Z_{n+1}+1) \mid \mathcal{F}_n] \le \log(Z_n+1) \quad \text{a.s.}$$

where recall that $\mathcal{F}_n$ is the σ-algebra generated by $Z_0, \ldots, Z_n$. Now, the claim of the theorem follows by proposition 3.6, part 1). ∎

Proof of theorem 3.4. Let $f(x) = (\log x)^{-\alpha}$ for $x > 0$ and $\alpha \le s-1$. It is shown in [KER 86] that for $x + h \ge 3$, where $x \ge 3$ and a sufficiently large positive constant C^*,

$$f(x+h) \le f(x) + f'(x)h + \frac{1}{2}f''(x)h^2 + C^*\frac{|h|^{2+\delta}}{(\log x)^{1+\alpha}x^{2+\delta}}$$
$$+\mathcal{I}(-\infty, -x/2](h).$$

Setting $x = Z_n + 3$ and $h = g(Z_n) + \xi_{n+1}$ and taking expectation of both sides, we obtain

$$E[f(Z_{n+1}+3) \mid Z_n = k] \le f(k+3) + f'(k+3)g(k)$$
$$+\frac{1}{2}f''(k+3)E[(g(k)+\xi_{n+1})^2]$$
$$+C^*E\left[\frac{|g(k)+\xi_{n+1}|^{2+\delta}}{(\log(k+3))^{1+\alpha}(k+3)^{2+\delta}}\right.$$

$$+\mathcal{I}\left(-\infty, -(k+3)/2\right]\left(g(k)+\xi_{n+1}\right)\Bigg].$$

Applying Chebyshev and C_r inequalities, we have for $k = 1, 2, \ldots$

$$\begin{aligned}
&E[f(Z_{n+1}+3) \mid Z_n = k] \\
&\leq f(k+3) + f'(k+3)g(k) + \frac{1}{2}f''(k+3)(g^2(k)+\ell_2(k)) \\
&\quad + C_1\left(\frac{g^{2+\delta}(k)}{(\log k)^{1+\alpha}k^{2+\delta}} + \frac{\ell_{2+\delta}(k)}{(\log k)^{1+\alpha}k^{2+\delta}}\right) + C_2\frac{\ell_{2+\delta}(k)}{k^{2+\delta}},
\end{aligned}$$

where C_1 and C_2 are positive constants. By (i) and (ii) and using a similar argument to that in the proof of theorem 3.3, we see that for large enough k and sufficiently small $\epsilon > 0$

$$\begin{aligned}
E[f(Z_{n+1}+3) \mid Z_n = k] \leq f(k+3) &- \frac{\alpha g(k)}{k(\log k)^{1+\alpha}}(1-\epsilon) \\
&+ \frac{\alpha \ell_2(k)}{2k^2(\log k)^{1+\alpha}}(1+\epsilon).
\end{aligned}$$

It follows from (iii) that, if Z_n is large enough, then

$$E[f(Z_{n+1}+3) \mid \mathcal{F}_n] \leq f(Z_n+3) \quad \text{a.s.},$$

where $\mathcal{F}_n$ is the σ-algebra generated by $Z_0, \ldots, Z_n$. To complete the proof, it is sufficient to apply proposition 3.6, part 2). ∎

Theorems 3.3 and 3.4 give an almost complete solution to the extinction problem in the critical case. Unfortunately, they do not cover all possibilities. For instance, the case $\lim_{k\to\infty} 2kg(k)/\ell_2(k) = 1$ is not included. In the following theorem, under additional assumptions, we complete our investigation of the extinction probability. The theorem includes limiting results concerning the survival probability $P(Z_n > 0)$ as well.

THEOREM 3.5.– *Let $\{Z_n\}_{n\geq 0}$ be a critical CB process, such that the control functions $\phi_0(k)$, $k = 1, 2, \ldots$ have infinitely divisible probability distributions. Assume also that:*

(i) $\tau_m(k) = 1 + \frac{c}{k}$ *for $c > 0$ and $k = 1, 2, \ldots$;*

(ii) $\ell_2(k) = \nu k + O(1)$ *for $\nu > 0$ and $k \to \infty$;*

(iii) $\sup_{k\geq 1}\left(g_k^{1/k}\right)'''(1) < \infty.$

Under these conditions, two statements hold.

1) If $2c/\nu \leq 1$*, then* $q_N = 1$ *for any* $N = 1, 2, \ldots$

2) As $n \to \infty$*, for some positive constants* k_1 *and* k_2*,*

$$P(Z_n > 0) \sim \begin{cases} \dfrac{k_1}{n^{1-(2c/\nu)}} & \textit{if} \quad \dfrac{2c}{\nu} < 1 \\ \\ \dfrac{k_2}{\log n} & \textit{if} \quad \dfrac{2c}{\nu} = 1. \end{cases}$$

COMMENT.– Comparing theorem 3.5 with the classical Kolmogorov's result $P(Z_n > 0) \sim 2/(\sigma^2 n)$ for critical GW processes $\{Z_n\}_{n\geq 0}$, we see that in critical CB processes $\{Z_n\}_{n\geq 0}$ the survival probability $P(Z_n > 0)$ decays to zero slower. This is due to the immigration component, which by assumption (i), enters each generation. The asymptotic results obtained here are consistent with those for critical GWM processes with migration stopped at zero given in Chapter 2.

The proof of theorem 3.5 is based on constructing appropriate bounds for the transition p.g.f.s and adapting results from [HÖP 85] and [HÖP 86] for population-size-dependent GW processes. Details of the proof can be found in [GON 05d].

In example 3.2, the processes satisfy either the assumptions of theorem 3.4 (non-extinction case) or theorem 3.5 (extinction case).

EXAMPLE 3.2.– Consider a CB process with $m = 1$ and $\sigma^2 < \infty$. Let $\phi_0(k)$ have a Poisson distribution with mean $k + c$ for $k = 1, 2, \ldots$ and $c > 0$; $\phi_0(0) \equiv 0$. Under these assumptions, we have for $k = 1, 2, \ldots$

$$\begin{aligned} \varepsilon(k) &= \nu^2(k) = k + c, \\ \ell_2(k) &= \nu^2(k) + \sigma^2 \varepsilon(k) \\ &= (\sigma^2 + 1)k + O(1) \qquad \text{as } k \to \infty. \end{aligned}$$

Both conditions (i) and (ii) in theorem 3.5 (with $\nu = \sigma^2 + 1$) hold. Condition (iii) is also satisfied since

$$g_k^{1/k}(s) = \exp\left\{-\frac{(k+c)(1-s)}{k}\right\}, \qquad 0 \leq s \leq 1.$$

Hence, by theorem 3.5, $\sigma^2 \geq 2c - 1$ implies $q_N = 1$.
On the other hand, we have as $k \to \infty$

$$\begin{aligned} E\left[\phi_0^2(k)\right] &= (k+c)^2 + (k+c) = O(k^2) \\ E\left[|\phi_0(k) - \varepsilon(k)|^4\right] &= 3(k+c)^2 + (k+c) = O(k^2). \end{aligned}$$

Therefore, applying Lyapunov inequality and the relations

$$E[\phi_0^{1+\delta/2}(k)] = O(k^{1+\delta/2}), \quad E[|\phi_0(k) - \varepsilon(k)|^{2+\delta}] = O(k^{1+\delta/2}), \qquad [3.14]$$

where $0 < \delta \leq 1$, we obtain (see also the comments after theorem 3.4) that $\ell_{2+\delta}(k) = O\left(k^{1+\delta/2}\right)$ as $k \to \infty$. Hence, as $k \to \infty$

$$\frac{\ell_{2+\delta}(k)}{\ell_2(k)k^\delta(\log k)^{-s}} \leq M\frac{k^{1+\delta/2}}{(\sigma^2+1)k + O(1)}(\log k)^s \to 0.$$

By theorem 3.4, we derive that $\sigma^2 < 2c - 1$ implies $q_N < 1$. □

The following simulations illustrate the situation with the above example. Let the process defined in example 3.2 have Poisson offspring distribution with $m = \sigma^2 = 1$ and $Z_0 = 1$. Therefore, if $c \leq 1$, then $q_1 = 1$ and if $c > 1$ then $q_1 < 1$. The extinction–expansion dichotomy yields that the sequence of process p.g.f.s $\{F_n(s)\}_{n\geq 0}$ converges as $n \to \infty$ to q_N for $0 \leq s < 1$. Figures 3.4 and 3.5 show the behavior of $F_n(s)$ and $P(Z_n = 0)$ in non-extinction ($c = 2.01$) and extinction ($c = 0.001$) cases, respectively.

3.3. Background and bibliographical notes

The CB processes with deterministic control functions $\phi(k)$ were introduced in [SEV 74], where asymptotic properties of the non-critical processes were obtained under the principal assumption $\phi(k) = \alpha(k + o(k))$ as $k \to \infty$, where α is a real positive number. A necessary and sufficient condition for the extinction of the process with superadditive deterministic control function was proved in [MOL 98]. Some of the results in [SEV 74] were extended to multiple CB processes with deterministic control functions in [ZUB 74] and to CB processes with random environments in [HOL 84].

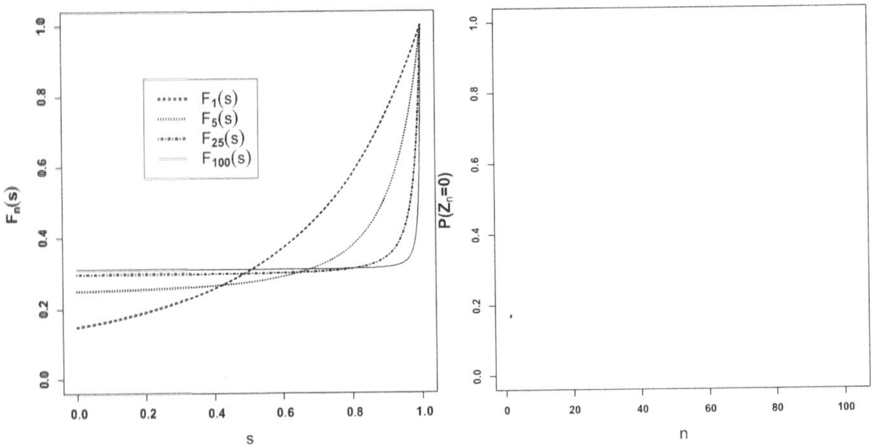

Figure 3.4. *Non-extinction case:* $F_n(s)$ *and* $P(Z_n = 0)$

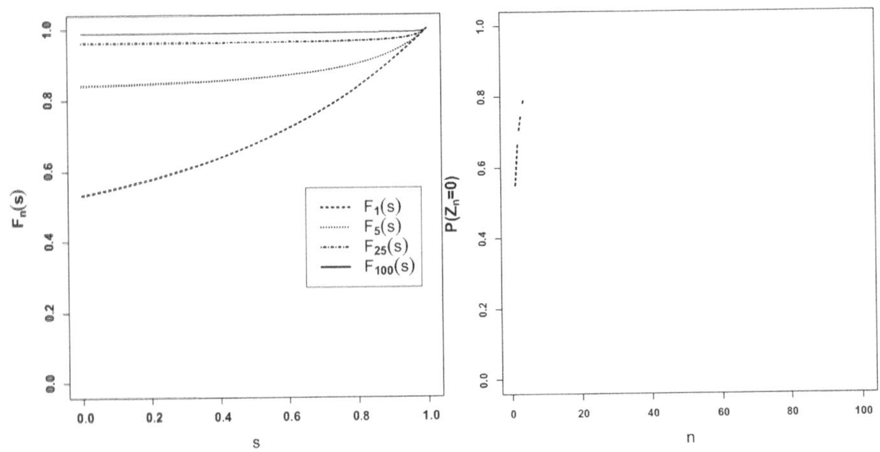

Figure 3.5. *Extinction case:* $F_n(s)$ *and* $P(Z_n = 0)$

The class of CB processes was substantially generalized in [YAN 75] by allowing the control functions $\{\phi_n(k)\}_{n\geq 0}$ to form a sequence of i.i.d. random variables, independent of the history of the process as defined in [3.1]. Later, this model was further extended to multiple random control functions in [YAN 89]. In these papers, as well as in [YAN 77, BRU 80] and [YAN 90], results regarding the probability of extinction were obtained under the condition $\phi_n(k) = \alpha_n(k + o(k))$ a.s. as $k \to \infty$, where $\{\alpha_n\}_{n\geq 0}$ is a sequence of i.i.d. random variables.

The classification of CB processes based on the asymptotic behavior of the mean growth rates $\tau_m(k)$ was established and further investigated in [GON 02, GON 04b, GON 04d, GON 04c, GON 05e]. Note that the concept of mean growth rate was introduced explicitly in the context of two-sex processes in [BRU 84]. A recent survey on critical CB processes is given in [YAN 15].

Over the years, a number of processes more or less overlapping with or closely related to the CB processes defined in this section were studied. Let us discuss some of them in the following.

CB processes with continuous state space were introduced in [ADK 95] and further studied in [RAH 07].

In [MAY 05], the idea of controlling a branching process by means of another branching process was introduced and explored.

Another interesting mathematical model, which may be seen as a particular type of CB process, is presented in [BRU 15].

The population-size-dependent processes, referred sometimes as controlled GW processes, were initially introduced in connection with some problems in chemistry and biology (see [LEV 68] and [LAB 72]). Papers on these processes, concerning conditions for extinction and asymptotic behavior on the non-extinction set, are [KÜS 85, HÖP 85, KLE 89a, KLE 89b] and references therein.

The two-sex processes were introduced in [DAL 68] with the purpose of modeling the probabilistic evolution of populations where females and males coexist and form couples. A good survey can be found in [MOL 10].

4

CB Processes: Limit Theorems

"Mathematics requires a small dose, not of genius, but of an imaginative freedom which, in a larger dose, would be insanity".
– Augus K. Rodgers

In this chapter, we study the asymptotic behavior of CB processes defined by [3.1]. Recall that, according to the classification in Chapter 3, a CB process is called **subcritical** if $\limsup_{k\to\infty} \tau_m(k) < 1$, **supercritical** if $\liminf_{k\to\infty} \tau_m(k) > 1$ and **critical** if

$$\liminf_{k\to\infty} \tau_m(k) \leq 1 \leq \limsup_{k\to\infty} \tau_m(k),$$

where $\tau_m(k) = k^{-1}E[Z_{n+1}|Z_n = k]$ for $k > 0$ are the mean growth rates.

4.1. Subcritical processes

As we have seen in Chapter 3, for the evolution of a subcritical CB process it matters whether or not 0 is an absorbing state. We will discuss limit theorems when **0 is not an absorbing state**. These results extend, for example, those for GWI processes with unrestricted immigration from Chapter 1 or those for GWM processes from Chapter 2. If, on the contrary, 0 is an absorbing state, then the extinction of a subcritical CB process is certain. In this case, one can expect the existence of a quasi-stationary limiting distribution given non-extinction similarly to that of a subcritical GW process. However, obtaining an analog for CB processes of Yaglom's theorem is still an open problem.

Recall from Chapter 3 that a CB process $\{Z_n\}_{n\geq 0}$ with 0 a non–absorbing state is an irreducible and aperiodic Markov chain if it satisfies the following conditions:

(A2): $p_0 > 0, \ \ p_0 + p_1 < 1$,

(A3): $P(\phi_0(k) > k) > 0$ for $k = 1, 2, \ldots$.

Also $P(\phi_0(0) > 0) > 0$ is equivalent to 0 is not an absorbing state.

The following result proves the existence of a stationary distribution.

THEOREM 4.1.– *Let $\{Z_n\}_{n\geq 0}$ be a subcritical CB process. If $P(\phi_0(0) > 0) > 0$ and (A2) and (A3) hold, then*

$$\lim_{n\to\infty} Z_n = Z \qquad \textit{in distribution},$$

where Z is a positive, finite and non-degenerate random variable.

Proof. Under the assumptions of the theorem, $\{Z_n\}_{n\geq 0}$ is an irreducible Markov chain with only one essential aperiodic class (see proposition 3.3). We will prove that $\{Z_n\}_{n\geq 0}$ is **positive recurrent** and therefore, according to Markov chain theory, $\{Z_n\}_{n\geq 0}$ converges in distribution to a positive, finite and non-degenerate random variable, whose probability law will be the stationary probability distribution.

First, we will prove that

$$\limsup_{n\to\infty} E[Z_n] < \infty. \qquad [4.1]$$

Indeed, $\limsup_{k\to\infty} \tau_m(k) < 1$ implies the existence of $k_0 > 0$ and $t < 1$ such that for $k = k_0 + 1, \ k_0 + 2, \ldots$ we have

$$\tau_m(k) = \frac{\varepsilon(k)}{k} m < t$$

and therefore for $k = 0, 1, \ldots$

$$m\varepsilon(k) \leq kt + mC, \qquad C := \max_{0\leq k\leq k_0} \varepsilon(k). \qquad [4.2]$$

By [4.2] we obtain for $n = 0, 1, \ldots$

$$\begin{aligned} E[Z_{n+1}] &= \sum_{k=0}^{\infty} E[Z_{n+1} \mid Z_n = k] P(Z_n = k) \qquad [4.3] \\ &= \sum_{k=0}^{\infty} m\varepsilon(k) P(Z_n = k) \end{aligned}$$

$$\leq tE[Z_n] + mC.$$

Iterating [4.3] and using that $Z_0 = N$, we have for $n = 0, 1, \ldots$

$$E[Z_{n+1}] \leq Nt^{n+1} + mC \sum_{k=0}^{n} t^k.$$

Now, since $t < 1$, we obtain

$$\limsup_{n\to\infty} E[Z_n] \leq \frac{mC}{1-t} < \infty,$$

which is [4.1]. Recall that $p_{ij}^{(n)} = P(Z_{m+n} = j \mid Z_m = i)$ for $n = 1, 2, \ldots$ and S denotes the state space of $\{Z_n\}_{n\geq 0}$. Assume, on the contrary, that $\{Z_n\}_{n\geq 0}$ is not a positive recurrent Markov chain. Then $\lim_{n\to\infty} p_{ij}^{(n)} = 0$ for all $i, j \in S$. However, since $Z_0 = N$, for every $l = 1, 2, \ldots$

$$\begin{aligned}\limsup_{n\to\infty} E[Z_n] &\geq \limsup_{n\to\infty} \sum_{i=l}^{\infty} i p_{Ni}^{(n)} \\ &\geq l \limsup_{n\to\infty} \left(1 - \sum_{i=0}^{l-1} p_{Ni}^{(n)}\right) \\ &= l,\end{aligned}$$

in contradiction with [4.1]. This completes the proof. ∎

Recall, from Chapter 3, that

$$g_k(s) = E\left[s^{\phi_0(k)}\right], \; f(s) = E\left[s^{X_{0,1}}\right], \quad \text{and} \quad F_n(s) = E\left[s^{Z_n}\right].$$

COROLLARY 4.1.– *Under the assumptions of theorem 4.1*

$$H(s) = E[g_Z(f(s))], \qquad 0 \leq s \leq 1, \tag{4.4}$$

where $H(s) := E[s^Z]$ is the p.g.f. of the limiting variable Z from theorem 4.1.

Proof. Theorem 4.1 implies $\lim_{n\to\infty} F_n(s) = H(s)$ uniformly for $0 \leq s \leq 1$. On the other hand, applying proposition 3.4, we obtain for $0 \leq s \leq 1$

$$\begin{aligned}
F_{n+1}(s) &= E\left[s^{Z_{n+1}}\right] \\
&= E\left[E\left[s^{Z_{n+1}}|Z_n\right]\right] \\
&= \sum_{k=0}^{\infty} h_k(s)P(Z_n = k) \\
&= \sum_{k=0}^{\infty} g_k(f(s))P(Z_n = k) \\
&\to \sum_{k=0}^{\infty} g_k(f(s))P(Z = k) \\
&= E[g_Z(f(s))].
\end{aligned}$$

∎

Differentiating both sides of [4.4] and evaluating the derivative at $s = 1$, we have

$$\begin{aligned}
E[Z] &= m\sum_{k=0}^{\infty} \varepsilon(k)P(Z = k) \\
&= mE[\varepsilon(Z)].
\end{aligned} \qquad [4.5]$$

Let us take a look at some subclasses of CB processes.

1) **GWI process.** If $\phi_n(k) = k + I_n$, where $\{I_n\}_{n\geq 0}$ is a sequence of non-negative integer-valued i.i.d. random variables, independent of $\{X_{n,j}\}_{n\geq 0, j\geq 1}$, then $\{Z_n\}_{n\geq 0}$ is a GWI process. Clearly $\varepsilon(k) = k + a$, where $0 < a := E[I_n] < \infty$. Consequently, $\tau_m(k) = m(1 + a/k)$ for $k \geq 1$, and thus the CB process is subcritical if and only if $m < 1$. Assuming $m < 1$ and using $g_k(f(s)) = f^k(s)\psi(f(s))$, where $\psi(s) := E\left[s^{I_0}\right]$, we have from [4.4]

$$H(s) = E\left[f^Z(s)\psi(f(s))\right], \qquad 0 \leq s \leq 1,$$

which leads to the equation

$$H(s) = \psi(f(s))H(f(s)), \qquad 0 \leq s \leq 1. \qquad [4.6]$$

Note that [4.6] is the generalized Schröder functional equation for the limiting p.g.f. of a subcritical GWI process (see [1.32] in Chapter 1). Moreover, [4.5] yields $E[Z] = mE[Z + a]$, which in turn implies $E[Z] = am/(1 - m)$.

2) **GWM process.** Consider a branching process with migration given by the control functions $\phi_n(k) = \max\{0, k + M_n\}$, where $\{M_n\}_{n \geq 0}$ are i.i.d. random variables such that

$$P(M_n = -1) = p, \quad P(M_n = 0) = q, \quad P(M_n = 1) = r,$$

where $p + q + r = 1$. This is a particular case of the GWM processes discussed in Chapter 2. Since $\varepsilon(0) = r$, $\varepsilon(k) = k + r - p$ for $k \geq 1$, we obtain $\tau_m(k) = m(1 + (r - p)/k)$. Therefore, the CB process is subcritical if and only if $m < 1$. Furthermore,

$$g_0(s) = 1 - r + rs \quad \text{and} \quad g_k(s) = (p + qs + rs^2)s^{k-1}, \quad k = 1, 2, \ldots$$

Hence, applying [4.4], we have

$$\begin{aligned} H(s) &= P(Z = 0)(1 - r + rf(s)) \\ &\quad + \left(p + qf(s) + rf^2(s)\right) \sum_{k=1}^{\infty} P(Z = k) f^{k-1}(s). \end{aligned}$$

It is not difficult to derive the following functional equation for $0 \leq s \leq 1$

$$H(s)f(s) = \left(p + qf(s) + rf^2(s)\right)H(f(s)) - p(1 - f(s))H(0). \quad [4.7]$$

Note that in cases of pure immigration of a single individual, namely when $r = 1$ and $p = q = 0$, equation [4.7] reduces to [4.6], with $\psi(s) = s$. Furthermore, taking into account

$$\begin{aligned} E[\varepsilon(Z)] &= \sum_{k=0}^{\infty} \varepsilon(k) P(Z = k) \\ &= E[Z] + r - p + pP(Z = 0), \end{aligned}$$

and [4.5], for the mean of the limiting variable we obtain

$$E[Z] = (r - p + pP(Z = 0))\frac{m}{1 - m}, \quad [4.8]$$

where $P(Z = 0) = H(0)$ can be calculated from [4.7] setting $s = 0$.

EXAMPLE 4.1.– As an illustration, consider a GWM process with $p = 0.2$, $q = 0.1$ and $r = 0.7$, with only one immigrant, and Poisson offspring distribution with $m = 0.8$. We have simulated 10,000 processes and run each of them for 10,000 generations (in order to guarantee convergence to the stationary distribution). A histogram of the probability mass function of the limiting variable Z is presented in Figure 4.1. Equation [4.8] with $P(Z = 0) = 0.297$ (calculated by simulation) yields $E[Z] = 2.2376$. On the other hand, the estimate, from the simulated data, of $E[Z]$ equals 2.2316.

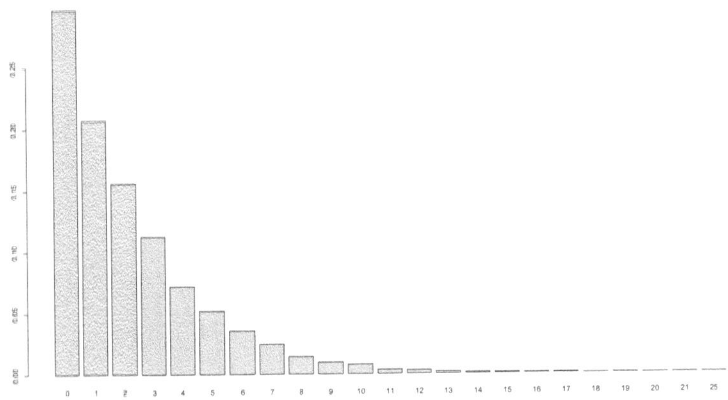

Figure 4.1. *Histogram of the limiting p.m.f. of a subcritical GWM process*

□

4.2. Critical processes

In this section, we will study CB processes satisfying condition *(A1)* (see proposition 3.2) in the critical case, i.e. when

$$\liminf_{k\to\infty} \tau_m(k) \leq 1 \leq \limsup_{k\to\infty} \tau_m(k).$$

The probability of extinction of a critical CB process was studied in detail in Chapter 3. Here, we will focus our attention on the limiting behavior of the process (in order to guarantee convergence to the stationary distribution) in both cases when the extinction is certain and when there is a positive non-extinction probability.

4.2.1. *Extinction is not certain*

Let us start with **processes having positive non-extinction probability**, i.e. $P(Z_n \to \infty) > 0$. Recall from Chapter 3 that $\{Z_n\}_{n\geq 0}$ satisfies the stochastic difference equation

$$Z_{n+1} = Z_n + g(Z_n) + Z_{n+1} - E[Z_{n+1} \mid Z_n], \qquad [4.9]$$

where $g(Z_n) = E[Z_{n+1} - Z_n \mid Z_n]$. Assume the mean growth rates satisfy for $0 < \alpha < 1$ and $c > 0$

$$\begin{aligned} \tau_m(k) &= \frac{1}{k} E[Z_{n+1} \mid Z_n = k] \\ &= 1 + \frac{c}{k^{1-\alpha}} + o\left(\frac{1}{k^{1-\alpha}}\right), \qquad \tau_m(k) > 1. \end{aligned}$$

Consequently, as $k \to \infty$

$$\begin{aligned} g(k) &:= E[Z_{n+1} \mid Z_n = k] - k \\ &= k(\tau_m(k) - 1) \\ &= ck^\alpha + o(k^\alpha). \end{aligned}$$

We extend $g(\cdot)$ to a twice differentiable function on $\mathbf{R}$ given for $0 < \alpha < 1$ by

$$g(x) = \begin{cases} cx^\alpha + o(x^\alpha) & x \geq 0, \quad x \to \infty \\ 0 & \text{otherwise.} \end{cases} \qquad [4.10]$$

Denote by $\{a_n\}_{n\geq 0}$ the solution of the deterministic recursive equation

$$a_0 = 1, \qquad a_{n+1} = a_n + g(a_n), \quad n = 0, 1, \ldots$$

Also recall from Chapter 3 that for $\zeta > 0$

$$\ell_\zeta(k) = E[|Z_{n+1} - E[Z_{n+1}|Z_n]|^\zeta \mid Z_n = k], \qquad k = 1, 2, \ldots$$

Under some assumptions on the mean growth rates $\tau_m(k)$ and the deviation moments $\ell_\zeta(k)$, we obtain theorem 4.2.

THEOREM 4.2.– *Let $\{Z_n\}_{n\geq 0}$ be a critical CB process. Let $0 < \alpha < 1$ and $c > 0$. Assume that the following relations are satisfied*

(i) As $k \to \infty$

$$\tau_m(k) = 1 + \frac{c}{k^{1-\alpha}} + o\left(\frac{1}{k^{1-\alpha}}\right), \qquad \tau_m(k) > 1.$$

(ii) For $\beta \leq \alpha + 1$ *and* $\nu > 0$

$$\ell_2(k) = \nu k^{\beta} + o(k^{\beta}), \qquad k \to \infty.$$

(iii) For some $\delta > 0$

$$\ell_{2+\delta}(k) = O((\ell_2(k))^{1+\delta/2}), \qquad k \to \infty.$$

Under these assumptions, the following limit results hold.

1) If $\beta = \alpha + 1$ *and* $2c/\nu > 1$, *then for any real number* z

$$\lim_{n\to\infty} P\left(\frac{1}{n} Z_n^{1-\alpha} \leq z \mid Z_r \to \infty\right) = \Gamma_{a,b}(z),$$

where $\Gamma_{a,b}(\cdot)$ *is the Gamma c.d.f. with parameters* $a = (2c - \nu\alpha)/(\nu - \nu\alpha)$ *and* $b = \nu(1-\alpha)^2/2$.

2) Let $\beta < \alpha + 1$.

(2.1) If $\beta < 3\alpha - 1$, *on* $\{Z_r \to \infty\}$,

$$\lim_{n\to\infty} \frac{Z_n}{a_n} = 1 \quad a.s. \quad and \quad \frac{Z_n - a_n}{g(a_n)} \text{ converges a.s.}$$

(2.2) If $\beta \geq 3\alpha - 1$, *on* $\{Z_r \to \infty\}$,

$$\lim_{n\to\infty} \frac{Z_n}{a_n} = 1 \qquad \text{in probability}$$

and for any real number z,

$$\lim_{n\to\infty} P\left(\frac{1}{\sqrt{\Delta_n}} \frac{Z_n - a_n}{g(a_n)} \leq z \mid Z_r \to \infty\right) = \Phi(z),$$

where Φ is the c.d.f. of the standard Normal distribution and

$$\Delta_n = \begin{cases} \dfrac{c_1(1-\alpha)^d}{d} n^d & \text{if } \beta > 3\alpha - 1, \\ \\ c_1 \log n & \text{if } \beta = 3\alpha - 1, \end{cases}$$

where $d = (\beta - 3\alpha + 1)/(1-\alpha)$ *and* $c_1 = \nu c^{(\beta-2)/(1-\alpha)}/(1-\alpha)$.

Let us verify that the assumptions of theorem 4.2 imply those of theorem 3.4 and therefore, the process $\{Z_n\}_{n\geq 0}$ has a **positive non-extinction probability**, i.e. $P(Z_n \to \infty) > 0$. Indeed, obviously (i) yields $\tau_m(k) = 1 + o(1)$. Furthermore, [4.10] and (ii) imply

$$\lim_{k\to\infty} \frac{2kg(k)}{\ell_2(k)} = \begin{cases} \frac{2c}{\nu} & \text{if } \beta = \alpha + 1, \\ \\ \infty & \text{otherwise}. \end{cases}$$

Finally, using (ii) and (iii), we obtain

$$\begin{aligned} \frac{\ell_{2+\delta}(k)}{\ell_2(k)k^\delta(\log k)^{-s}} &\leq M \frac{(\ell_2(k))^{\delta/2}}{k^\delta}(\log k)^s \\ &= M \frac{\nu^{\delta/2}k^{\beta\delta/2} + o(k^{\beta\delta/2})}{k^\delta}(\log k)^s \\ &= M\nu^{\delta/2}k^{\delta(\beta-2)/2}(1+o(1))(\log k)^s \\ &\to 0 \quad \text{as} \quad k \to \infty, \end{aligned}$$

and therefore (see comment to theorem 3.4)

$$\ell_{2+\delta}(k) = o\left(\frac{g(k)k^{1+\delta}}{(\log k)^s}\right).$$

Note that, if $\beta > \alpha + 1$ or if $\beta = \alpha + 1$ and $2c/\nu < 1$, then the probability of extinction is one (see theorem 3.3).

Figure 4.2 summarizes the asymptotic results in the critical case. Note that, given non-extinction, under broad assumptions, Z_n (properly normalized) is asymptotically normally distributed. However, under the special circumstances when the asymptotic behavior of $g(k)$ (the mean difference between two consecutive population sizes) and the conditional variance $\ell_2(k)$ is balanced, the limit is Gamma distributed.

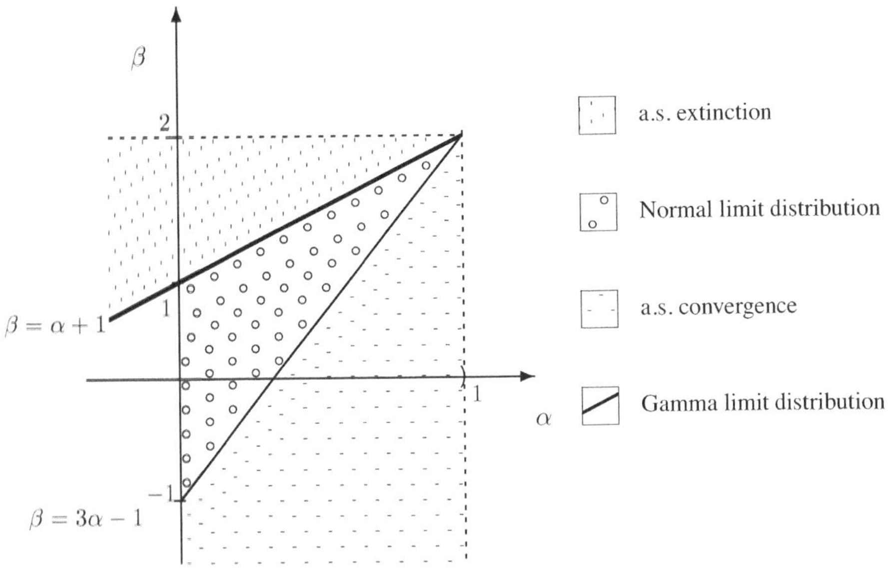

Figure 4.2. *Regions with different asymptotic behavior in the critical case. Reprinted by permission of the Applied Probability Trust. First published in [GON 05]. Copyright © Applied Probability Trust 2005*

Proof of theorem 4.2. We will need the function

$$G(x) := \int_1^x \frac{1}{g(z)} dz, \quad x \geq 1.$$

Since, by definition, $g(x) = cx^\alpha + o(x^\alpha)$, we have as $x \to \infty$

$$g'(x) = \frac{c\alpha}{x^{1-\alpha}} + o\left(\frac{1}{x^{1-\alpha}}\right) \quad \text{and} \quad G(x) \sim \frac{x^{1-\alpha}}{c(1-\alpha)}. \qquad [4.11]$$

Proof of part 1). It follows from [4.11] that

$$\lim_{x\to\infty} g'(x)G(x) = \frac{1}{1-\alpha} - 1,$$

and, referring to (ii), we obtain

$$\lim_{k\to\infty} \frac{\ell_2(k)}{g^2(k)G(k)} = \frac{\nu(1-\alpha)}{c} > 0.$$

Denote $\lambda := 1/(1-\alpha)$ and $\gamma := \nu(1-\alpha)/c$. Since $2c/\nu > 1$ we have $\gamma\lambda < 2$. Therefore, the assumptions in theorem A3.2 in Appendix 3 hold and thus

$$\lim_{n\to\infty} P\left(\frac{G(Z_n)}{n} \le z \mid Z_r \to \infty\right) = \Gamma_{a,b}(z), \qquad [4.12]$$

where $a = 2/\gamma - \lambda + 1$ and $b = \gamma/2$. Now, applying [4.11], we have on $\{Z_r \to \infty\}$

$$\lim_{n\to\infty} \frac{\lambda Z_n^{1-\alpha}}{cG(Z_n)} = 1 \quad \text{a.s.} \qquad [4.13]$$

Finally, taking into account [4.12] and [4.13], Slutsky's theorem (see theorem A.2.1 in Appendix 2), and using well-known properties of Gamma distribution, we complete the proof of claim 1).

Proof of part 2). We will use theorem A3.1 in Appendix 3. Let us verify that its assumptions hold. Given $g(Z_n) \neq 0$, equation [3.11] can be written in the form

$$Z_{n+1} = Z_n + g(Z_n)(1 + \eta_{n+1}), \quad n = 0, 1, \ldots, \qquad [4.14]$$

where

$$\eta_{n+1} := \frac{Z_{n+1} - E[Z_{n+1}|Z_n]}{g(Z_n)}.$$

Obviously, $\{\eta_n\}_{n\ge 1}$ are zero-mean, square-integrable martingale differences with respect to the family of σ–algebras $\mathcal{F}_n = \sigma(Z_0, \ldots, Z_n)$, $n = 0, 1, \ldots,$ and

$$E[\eta_{n+1}^2 \mid Z_n = k] = \frac{\ell_2(k)}{g^2(k)}, \quad k = 1, 2, \ldots$$

Assumption (iii) implies

$$E[|\eta_{n+1}|^{2+\delta} \mid Z_n = k] = O\left((\varphi^2(k))^{1+\delta/2}\right), \qquad [4.15]$$

where $\varphi^2(k) := E[\eta_{n+1}^2 \mid Z_n = k]$. Consider the function

$$\psi(x) := \int_1^x \frac{\varphi^2(z)}{g(z)} dz, \quad x \geq 1.$$

Assumption (ii) and $g(x) = cx^\alpha + o(x^\alpha)$ yield

$$\frac{\varphi^2(x)}{g(x)} \sim \frac{\nu}{c^3} x^{\beta - 3\alpha}, \qquad x \to \infty,$$

and therefore, as $x \to \infty$,

$$\psi(x) \sim \begin{cases} \dfrac{\nu\left(x^{\beta-3\alpha+1} - 1\right)}{c^3(\beta - 3\alpha + 1)} & \text{if } \beta \neq 3\alpha - 1, \\ \dfrac{\nu}{c^3} \log x & \text{if } \beta = 3\alpha - 1. \end{cases} \qquad [4.16]$$

For the inverse function G^{-1} of G, we have, as $x \to \infty$

$$G^{-1}(x) \sim (c(1-\alpha)x)^{1/(1-\alpha)}.$$

Consequently, if $\widehat{\psi}(x) := \psi(G^{-1}(x))$, then, as $x \to \infty$,

$$\widehat{\psi}(x) \sim \begin{cases} \dfrac{\nu}{c^3(\beta - 3\alpha + 1)} \left((c(1-\alpha)x)^{\frac{\beta-3\alpha+1}{1-\alpha}} - 1\right) & \text{if } \beta \neq 3\alpha - 1, \\ \dfrac{\nu}{c^3(1-\alpha)} \log x & \text{if } \beta = 3\alpha - 1. \end{cases} \qquad [4.17]$$

Let us summarize our findings so far.

(A) $g(x)$ is positive, twice differentiable on $[0, +\infty)$, asymptotically concave, and $g'(x)$ is asymptotically convex as $x \to \infty$.

(B) $\varphi^2(x)$ is positive and continuously differentiable. Furthermore,

$$\int_1^x \frac{\varphi^2(G^{-1}(z))}{z^2} dz \sim \frac{\nu}{c^2}(c(1-\alpha))^{\frac{\beta-2\alpha}{1-\alpha}} \int_1^x z^{\frac{\beta-2\alpha}{1-\alpha}-2} dz,$$

and, since $\beta < \alpha + 1$, the integral on the right-hand side converges as $x \to \infty$. Moreover, as $x \to \infty$

$$\varphi^2(G^{-1}(x)) \sim \frac{\nu}{c^2}(c(1-\alpha)x)^{\frac{\beta-2\alpha}{1-\alpha}},$$

so it is asymptotically either concave or convex.

(C) If $\beta \geq 3\alpha - 1$, then [4.16] implies $\lim_{x\to\infty} \psi(x) = \infty$ and as $x \to \infty$

$$\left|\frac{g''(x)g(x)}{\varphi^2(x)}\right| = O\left(\frac{1}{x^{\beta+2-4\alpha}}\right),$$

which is asymptotically decreasing because $\beta + 2 - 4\alpha > 0$. On the other hand, if $\beta < 3\alpha - 1$, then [4.16] implies $\lim_{x\to\infty} \psi(x) < \infty$ and, as $x \to \infty$,

$$|g''(x)g(x)| = O\left(\frac{1}{x^{2(1-\alpha)}}\right),$$

which is asymptotically decreasing since $1 - \alpha > 0$.

(D) If $\beta \geq 3\alpha - 1$, then, as $x \to \infty$,

$$g'(x)\psi^{1/2}(x) = \begin{cases} O\left(x^{\frac{\beta-\alpha-1}{2}}\right) & \text{if } \beta > 3\alpha - 1, \\ O\left(\frac{(\log x)^{1/2}}{x^{1-\alpha}}\right) & \text{if } \beta = 3\alpha - 1. \end{cases}$$

If $\beta < 3\alpha - 1$, then as $x \to \infty$

$$\begin{aligned} g'(x)\psi^{1/2}(x) &\sim \alpha\left(\frac{\nu(1 - x^{\beta-3\alpha+1})}{c(3\alpha - 1 - \beta)}\right)^{1/2} \frac{1}{x^{1-\alpha}} \\ &\to 0. \end{aligned}$$

Equations [4.15], [4.17] and (A)–(D) imply that the assumptions in theorem A3.1 of Appendix 3 hold. Using the second part of this theorem, we obtain the result in (2.1). The first part of theorem A3.1 in Appendix 3 yields that for $\beta \geq 3\alpha - 1$

$$\lim_{n\to\infty} \frac{Z_n}{a_n} = 1 \quad \text{in probability on} \quad \{Z_r \to \infty\},$$

and for all real z,

$$\lim_{n\to\infty} P\left(\frac{1}{\sqrt{\widehat{\psi}(n)}} \frac{Z_n - a_n}{g(a_n)} \leq z \mid Z_r \to \infty\right) = \Phi(z),$$

where $\Phi(\cdot)$ is the c.d.f. of the standard Normal distribution. Finally, since, as $n \to \infty$,

$$\widehat{\psi}(n) \sim \begin{cases} \frac{\nu}{\beta - 3\alpha + 1} c^{\frac{\beta-2}{1-\alpha}} ((1-\alpha)n)^{\frac{\beta-3\alpha+1}{1-\alpha}} & \text{if } \beta > 3\alpha - 1, \\ \frac{\nu}{c^3(1-\alpha)} \log n & \text{if } \beta = 3\alpha - 1, \end{cases}$$

applying Slutsky's theorem we complete the proof of (2.2). ∎

COROLLARY 4.2.–

1) Under the assumptions in theorem 4.2, part 1), for any real number z,

$$\text{(1.1)} \lim_{n\to\infty} P\left(\frac{Z_n^{1-\alpha}}{n} \leq z\right) = q_N \mathcal{I}[0,+\infty)(z) + (1-q_N)\Gamma_{a,b}(z),$$

$$\text{(1.2)} \lim_{n\to\infty} P\left(\frac{Z_n^{1-\alpha}}{n} \leq z \middle| Z_n > 0\right) = \Gamma_{a,b}(z),$$

where $a = (2c - \nu\alpha)/(\nu - \nu\alpha)$ *and* $b = \nu(1-\alpha)^2/2$.

2) Under the assumptions in theorem 4.2, part 2), for any real number z,

$$\lim_{n\to\infty} P\left(\frac{1}{\sqrt{\Delta_n}} \frac{Z_n - a_n}{g(a_n)} \leq z \mid Z_n > 0\right) = \Phi(z).$$

Proof. 1) Set, for simplicity, $Y_n := Z_n^{1-\alpha}/n$. Applying theorem 4.2, part 1) and the extinction-expansion dichotomy, we obtain for $z > 0$

$$\begin{aligned}\lim_{n\to\infty} P(Y_n \leq z) &= \lim_{n\to\infty} P(Y_n \leq z \mid Z_r \to 0)P(Z_r \to 0)\\ &\quad + \lim_{n\to\infty} P(Y_n \leq z \mid Z_r \to \infty)P(Z_r \to \infty)\\ &= P(Z_r \to 0)\\ &\quad + \lim_{n\to\infty} P(Y_n \leq z \mid Z_r \to \infty)P(Z_r \to \infty)\\ &= q_N + (1-q_N)\Gamma_{a,b}(z),\end{aligned}$$

i.e. (1.1) holds. To prove (1.2), we observe that

$$\begin{aligned}P(Y_n \leq z, Z_n > 0) &= P(Y_n \leq z, Z_n > 0, Z_r \to \infty)\\ &\quad + P(Y_n \leq z, Z_n > 0, Z_r \to 0)\\ &\leq P(Y_n \leq z, Z_r \to \infty) + P(Z_n > 0, Z_r \to 0).\end{aligned}$$

Hence,

$$\lim_{n\to\infty} P(Y_n \leq z, Z_n > 0) = \lim_{n\to\infty} P(Y_n \leq z, Z_r \to \infty),$$

and therefore,

$$\lim_{n\to\infty} P(Y_n \le z \mid Z_n > 0) = \frac{\lim_{n\to\infty} P(Y_n \le z,\ Z_n > 0)}{\lim_{n\to\infty} P(Z_n > 0)}$$
$$= \frac{\lim_{n\to\infty} P(Y_n \le z,\ Z_r \to \infty)}{P(Z_r \to \infty)}$$
$$= \lim_{n\to\infty} P(Y_n \le z \mid Z_r \to \infty),$$

which, in view of theorem 4.2, part 1), implies (1.2).

2) The proof is similar to that of (1.2), with $Y_n := \Delta_n^{-1/2}(Z_n - a_n)/g(a_n)$ and applying theorem 4.2, part 2) instead of theorem 4.2, part 1). ∎

Let us show that

$$a_n \sim (c(1-\alpha)n)^{1/(1-\alpha)}, \qquad n \to \infty. \qquad [4.18]$$

By the definition of $\{a_n\}_{n\ge 0}$, we have $\lim_{n\to\infty} a_n = \infty$. Taking into account [4.10] and lemma A3.1 in Appendix 3, we have as $n \to \infty$

$$a_{n+1}^{1-\alpha} = a_n^{1-\alpha}\left(1 + \frac{a_{n+1} - a_n}{a_n}\right)^{1-\alpha}$$
$$= a_n^{1-\alpha} + (1-\alpha)\frac{g(a_n)}{a_n^{\alpha}} + O\left(\frac{g^2(a_n)}{a_n^{1+\alpha}}\right)$$
$$= a_n^{1-\alpha} + (1-\alpha)c + o(1).$$

Hence,

$$a_n^{1-\alpha} = c(1-\alpha)n + o(n), \qquad n \to \infty,$$

which proves [4.18].

COMMENT.– Regarding assumption (iii) of theorem 4.2, see part (b) of the comment after theorem 3.4.

4.2.2. *Extinction is certain*

Under the conditions of theorem 3.5, **the extinction is certain**. We will study the limiting distribution of Z_n/n given $\{Z_n > 0\}$.

THEOREM 4.3.– *Under the assumptions of theorem 3.5, if, additionally,* $2c/\nu \leq 1$, *then for every real number* z,

$$\lim_{n\to\infty} P\left(\frac{Z_n}{n} \leq z \mid Z_n > 0\right) = \Gamma_{1,\nu/2}(z).$$

Similarly to theorem 3.5, the proof of theorem 4.3 makes use of results for population-size-dependent GW processes from [HÖP 85] and [HÖP 86]. A detailed proof can be found in [GON 05d].

COMMENT.– In theorem 4.3, the first parameter of the limiting Gamma distribution does not depend on c and ν. This is in contrast to the case when the extinction is not certain.

There are similarities between theorem 4.3 and the classical theorem 1.5 for critical GW processes. In both the limit on the non-extinction paths is Exponential, in spite of the fact that the decay rates of the two processes are different (see theorem 3.5). The Gamma limit for GWI processes (see theorem 1.9) can also be recovered from theorem 4.3. An analogous result holds for critical branching processes with random migration when the immigration dominates emigration, see theorem 2.3.

EXAMPLE 4.2.– Revisiting example 3.2, consider a CB process with $m = 1$ and $0 < \sigma^2 < \infty$. Let $\phi_0(0) = 0$ and $\phi_0(k)$ have a Poisson distribution with mean $k + c$ for $k = 1, 2, \ldots$ and $c > 0$. For the limiting distribution of Z_n/n given $Z_n > 0$, we have, by theorem 4.3, that if $\sigma^2 \geq 2c - 1$, then it is Gamma with parameters $a = 1$ and $b = (\sigma^2 + 1)/2$, whereas, by corollary 4.2 (1.2), if $\sigma^2 < 2c - 1$, then it is Gamma with parameters $a = 2c/(\sigma^2 + 1)$ and $b = (\sigma^2 + 1)/2$. Let us set $Z_0 = 1$ and Poisson offspring distribution. Then the limiting distribution of Z_n/n given $Z_n > 0$ is Gamma with parameters $a = 1$ and $b = 1$ when $c \leq 1$ and Gamma with parameters $a = c$ and $b = 1$ when $c > 1$. As an illustration, in Figure 4.3, we plot the Laplace transform of Z_n/n given $Z_n > 0$ for several values of n along with the theoretical limit (solid line), setting $c = 0.001$ (hence $q_1 = 1$) and $c = 2.01$ (hence $q_1 < 1$). □

4.2.3. *Feller diffusion approximation*

In this section, we discuss Feller diffusion approximation for an array of CB processes whose offspring and control means approach their critical values. The motivation for this study is twofold. From a theoretical viewpoint, obtaining functional weak-limit theorems for branching processes has been of

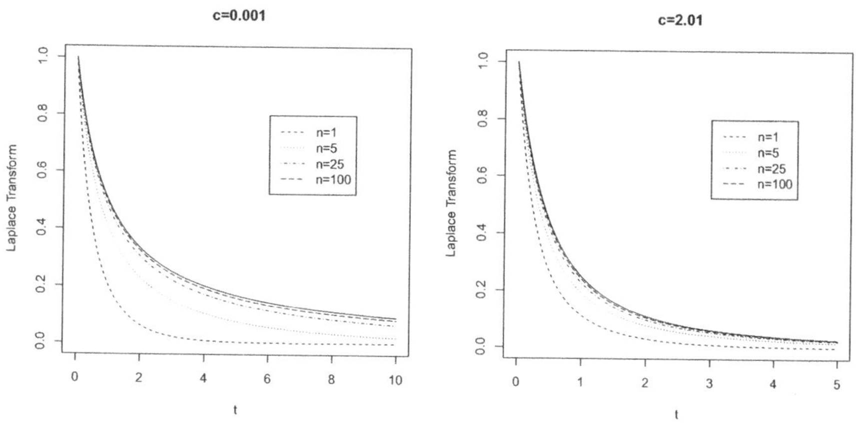

Figure 4.3. *Convergence of the Laplace transforms of Z_n/n given $Z_n > 0$*

interest for a long time. Here, we will present some recent advances. From a practical viewpoint, these limit theorems for sequences of branching processes are useful in studying asymptotic properties of certain parameter estimators, as we will see in Chapter 5.

Define an array of CB processes $\{Z_i^{(n)}\}_{i\geq 0}$ for $n = 1, 2, \ldots$ by

$$Z_{i+1}^{(n)} = \sum_{j=1}^{\phi_i^{(n)}\left(Z_i^{(n)}\right)} X_{i,j}^{(n)}, \quad i = 0, 1, \ldots; \; n = 1, 2, \ldots; \quad Z_0^{(n)} = N, \qquad [4.19]$$

where, for every fixed n, $\{X_{i,j}^{(n)}\}_{i\geq 0, j\geq 1}$ is a sequence of non-negative integer-valued i.i.d. random variables with mean m_n and variance σ_n^2, and $\{\phi_i^{(n)}(k)\}_{i\geq 0, k\geq 0}$ are independent, non-negative integer-valued random variables with means $\varepsilon_n(k)$ and variances $\nu_n^2(k)$ for every $k \geq 0$. Also assume that $\{X_{i,j}^{(n)}\}$ and $\{\phi_i^{(n)}(k)\}$ are independent for every n.

Introduce the sequence of random functions $\{W_n\}_{n\geq 1}$ by

$$W_n(t) := \frac{Z_{[nt]}^{(n)}}{n}, \; t \geq 0, \; n = 1, 2, \ldots,$$

where recall that $[x]$ denotes the integer part of x. $\{W_n\}_{n\geq 1}$ are $D_{[0,\infty)}[0,\infty)$–valued random variables, where $D_{[0,\infty)}[0,\infty)$ is the space of non-negative functions on $[0,\infty)$ which are right continuous and have limits from the left. Denote by $C_c^{\infty}[0,\infty)$ the space of infinitely differentiable functions on $[0,\infty)$ with compact support. Let $\xrightarrow{\mathcal{D}}$ stand for convergence of random functions in the Skorokhod topology (also known as weak convergence). Furthermore, we need the following assumptions:

(H1) Let $0 < m < \infty$, $-\infty < \alpha < \infty$. Assume

$$m_n = m + \frac{\alpha}{n} + o\left(\frac{1}{n}\right), \qquad n \to \infty.$$

(H2) $$\lim_{n\to\infty} \sigma_n^2 = \sigma^2, \qquad 0 < \sigma^2 < \infty.$$

(H3) For any sequence $\{x_n\}_{n\geq 1}$, such that $\lim_{n\to\infty} x_n = x$ for $0 < x < \infty$ and for any $\epsilon > 0$, assume

$$\lim_{n\to\infty} E\left[|X_{0,1}^{(n)} - m_n|^2 \mathcal{I}\left\{|X_{0,1}^{(n)} - m_n| \geq \epsilon\sqrt{n x_n \sigma_n^2}\right\}\right] = 0.$$

(H4) $$\varepsilon_n(k) = \varepsilon(k) + f_n(k),$$

where $\lim_{n\to\infty} f_n(k) = 0$ uniformly in k.

(H5) Assume for $-\infty < \gamma < \infty$

$$\frac{\varepsilon(k)}{k} m = 1 + \frac{\gamma}{k} + o\left(\frac{1}{k}\right), \qquad k \to \infty.$$

(H6) $$\nu_n^2(k) = \beta_n k + g_n(k), \quad \lim_{k\to\infty} \frac{g_n(k)}{k} = 0 \quad \text{uniformly in } n,$$

and $\lim_{n\to\infty} \beta_n = \beta$ with $0 \leq \beta < \infty$.

It follows from (H1), (H4) and (H5) that the limit of the mean growth rates of the process $\{Z_i^{(n)}\}_{i\geq 0}$, as $k \to \infty$, equals m_n/m and converges to one as $n \to \infty$. Thus, although each process $\{Z_i^{(n)}\}_{i\geq 0}$ is not necessarily critical, we can say that in some sense the offspring and control means approach their critical values. Processes with these properties are often called **close-to-critical**.

Using operator semigroup convergence results, the following weak convergence theorem for the sequence of random functions $\{W_n\}_{n\geq 0}$ can be proved (see [GON 12a]).

THEOREM 4.4.– *Assume (H1)–(H6) and let* $\beta = 0$. *Then* $W_n \xrightarrow{\mathcal{D}} W_\alpha$ *as* $n \to \infty$, *where* W_α *is the diffusion process with generator*

$$A_\alpha f(x) = \left(\gamma + \frac{\alpha x}{m}\right) f'(x) + \frac{\sigma^2 x}{2m} f''(x), \quad f \in C_c^\infty[0, \infty). \qquad [4.20]$$

The process W_α *is the (unique) solution of the stochastic differential equation*

$$dW_\alpha(t) = \left(\gamma + \frac{\alpha}{m} W_\alpha(t)\right) dt + \left(\frac{\sigma^2}{m} W_\alpha(t)\right)^{1/2} dB(t), \quad t \geq 0,$$

where B *is a standard Wiener process.*

COMMENT.– (i) The theorem still holds if $\{Z_0^{(n)}\}_{n\geq 1}$ in [4.19] is a sequence of random variables such that $Z_0^{(n)}/n$ converges in distribution as $n \to \infty$.

(ii) It can be proven that theorem 4.4 still holds after replacing the assumptions (H4)–(H5) by

$$m\frac{\varepsilon_n(k)}{k} = 1 + \frac{\gamma_n}{k} + h_n(k),$$

where $\lim_{k\to\infty} h_n(k)/k = 0$ and $\lim_{n\to\infty} \gamma_n = \gamma$ with $-\infty < \gamma < \infty$.

It is not difficult to see that if (H1), (H2) and (H4)–(H6) hold and, in addition, $\beta = \sigma^2 = 0$, then the sequence $\{W_n\}_{n\geq 1}$ converges weakly to a diffusion process with generator

$$A_\alpha f(x) = \left(\gamma + \frac{\alpha x}{m}\right) f'(x), \qquad f \in C_c^\infty[0, \infty).$$

In this case, the limiting process W_α is a deterministic function given by

$$W_\alpha(t) = \frac{\gamma m}{\alpha}\left(e^{\alpha t/m} - 1\right), \quad t \geq 0,$$

which is a solution of the (non-random) differential equation

$$dW_\alpha(t) = \left(\gamma + \frac{\alpha}{m} W_\alpha(t)\right) dt, \quad t \geq 0.$$

If $\gamma = 0$, then $W_\alpha(t) = 0$ for all $t \geq 0$. Note that the assumption (H3) is not needed as long as $\lim_{n\to\infty} \sigma_n^2 = 0$ (for details, see [GON 12a]).

Let us consider some subclasses of CB processes.

1) **GWI process**. GWI processes are an important subclass of the CB processes, as it was mentioned before. They are given by $\phi_i^{(n)}(k) = k + I_i^{(n)}$, where for every n, $\{I_i^{(n)}\}_{i\geq 0}$ is a sequence of non-negative integer-valued i.i.d. random variables, independent of $\{X_{i,j}^{(n)}\}_{i\geq 0, j\geq 1}$. For a sequence of critical GWI processes, assuming that (H1)–(H6) hold and setting $m = 1$, $\beta = 0$, $\sigma^2 = \sigma_0^2$ and $\gamma = \lambda_0$, theorem 4.4 implies the weak convergence to a diffusion process with generator

$$A_\alpha f(x) = (\lambda_0 + \alpha x) f'(x) + \frac{\sigma_0^2 x}{2} f''(x), \quad f \in C_c^\infty[0,\infty).$$

2) **CB process in varying environment.** A natural modification of CB processes is one allowing the reproductive ability (offspring distribution) of the population to vary across generations. The so-called CB processes in varying environment (see [GON 15]) are defined with the same hypotheses on the control functions as ordinary CB processes. However, the offspring variables $\{X_{n,i}\}_{n\geq 0, i\geq 1}$, which are still independent, do not have to be identically distributed with respect to n. Denote the mean and variance of $X_{n,i}, i = 1, 2, \ldots$ by m_n and σ_n^2, respectively. Applying theorem 4.4, we obtain the following result.

THEOREM 4.5.– *Let $\{Z_n\}_{n\geq 0}$ be a CB process in varying environment. Assume the following conditions hold.*

(i) For $0 < m < \infty$ and $-\infty < \alpha < \infty$

$$m_n = m + \frac{\alpha}{n} + o\left(\frac{1}{n}\right), \qquad n \to \infty.$$

(ii) $\lim_{n\to\infty} \sigma_n^2 = \sigma^2$, where $0 < \sigma^2 < \infty$.

(iii) For $\{x_n\}_{n\geq 1}$, such that $\lim_{n\to\infty} x_n = x$, where $0 < x < \infty$, and for every $\epsilon > 0$,

$$\lim_{n\to\infty} \frac{1}{\sigma_n^2} E\left[|X_{0,1} - m_n|^2 \mathcal{I}\left\{|X_{0,1} - m_n| \geq \epsilon\sqrt{n x_n \sigma_n^2}\right\}\right] = 0.$$

(iv) For $-\infty < \gamma < \infty$

$$\frac{\varepsilon(k)}{k}m = 1 + \frac{\gamma}{k} + o\left(\frac{1}{k}\right), \qquad k \to \infty.$$

(v) $$\lim_{k\to\infty} \frac{\nu^2(k)}{k} = \beta, \qquad 0 \le \beta < \infty.$$

Then $W_n \xrightarrow{\mathcal{D}} W_\alpha$*, as* $n \to \infty$*, where* $W_n(t) = Z_{[nt]}/n$*,* $t \ge 0$*, and* W_α *is a non-negative diffusion process with generator*

$$A_\alpha f(x) = \left(\gamma + \frac{\alpha x}{m}\right) f'(x) + \frac{\sigma^2 x}{2m} f''(x), \qquad f \in C_c^\infty[0,\infty).$$

It is well-known (see [IKE 89, p. 236]) that the diffusion process W_α satisfies

$$E\left[e^{-sW_\alpha(t)} \mid W_\alpha(0) = x\right] = \left(\frac{\sigma^2 s}{2\alpha}\left(e^{\alpha t/m} - 1\right) + 1\right)^{-2\gamma m/\tau^2} \cdot \exp\left(-\frac{sxe^{\alpha t/m}}{\sigma^2 s(e^{\alpha t/m} - 1)/(2\alpha) + 1}\right).$$

4.3. Supercritical processes

In this section, we continue our discussion of CB processes satisfying (A1) (and thus the extinction–expansion dichotomy holds). More specifically, we focus our attention on the supercritical case, i.e. when $\liminf_{k\to\infty} \tau_m(k) > 1$.

Assume $\{\varepsilon(k)/k\}_{k\ge1}$ converges to a finite limit τ and hence $\lim_{k\to\infty} \tau_m(k) = \tau m$. We will begin with proving the almost sure convergence of the process, suitably normalized, to a non-negative and finite random variable. Define $W_n := Z_n/(\tau m)^n$ for $n \ge 0$ (W_n is not to be confused with the random function with the same notation in section 4.2.3).

THEOREM 4.6.– *Let* $\{Z_n\}_{n\ge0}$ *be a CB process satisfying one of the following two conditions:*

(i) $\{\varepsilon(k)/k\}_{k\ge1}$ *is non-decreasing and convergent,*

(ii) $\{\varepsilon(k)/k\}_{k\geq 1}$ *is non-increasing, convergent and* $\{E[W_n]\}_{n\geq 0}$ *is bounded.*

Then, $\lim_{n\to\infty} W_n = W$ *a.s., where* W *is a non–negative and finite random variable.*

Proof. Let $\mathcal{F}_n := \sigma(Z_0, Z_1, \ldots, Z_n)$ for $n \geq 0$. We will prove that if *(i)* holds, then $\{W_n, \mathcal{F}_n\}_{n\geq 0}$ is a non-negative supermartingale. Indeed,

$$
\begin{aligned}
E[W_{n+1} \mid \mathcal{F}_n] &= \frac{E[Z_{n+1} \mid \mathcal{F}_n]}{(\tau m)^{n+1}} \\
&= \frac{\varepsilon(Z_n)m}{(\tau m)^{n+1}} \\
&\leq \frac{Z_n \tau m}{(\tau m)^{n+1}} \\
&= W_n \quad \text{a.s.}
\end{aligned}
$$

Hence, the supermartingale convergence theorem (see corollary 1 of theorem A1.1 in Appendix 1) implies that there exists a random variable W such that $\lim_{n\to\infty} W_n = W$ a.s. and $P(0 \leq W < \infty) = 1$.

Next, assuming that *(ii)* holds, we will prove that $\{W_n, \mathcal{F}_n\}_{n\geq 0}$ is a bounded submartingale in L^1. Indeed,

$$
\begin{aligned}
E[W_{n+1} \mid \mathcal{F}_n] &= \frac{\varepsilon(Z_n)m}{(\tau m)^{n+1}} \\
&\geq \frac{\tau m Z_n}{(\tau m)^{n+1}} \\
&= W_n \quad \text{a.s.}
\end{aligned}
$$

and also $\sup_{n\geq 0} E[|W_n|] = \sup_{n\geq 0} E[W_n] < \infty$. Now, the submartingale convergence theorem (see theorem A1.1 in Appendix 1) yields the theorem's assertion. ∎

The next theorem shows that $\{(\tau m)^n\}_{n\geq 0}$ is the only normalizing sequence leading to a geometric growth of the process.

THEOREM 4.7.– *Let* $\{Z_n\}_{n\geq 0}$ *be a CB process, such that:*

(i) $\lim_{k\to\infty} \frac{\varepsilon(k)}{k} = \tau < \infty, \qquad \tau m > 1,$

(ii) $\{\nu^2(k)/k\}_{k\geq 1}$ *is a bounded sequence.*

Then, on $\{Z_n \to \infty\}$,

$$\lim_{n\to\infty} \frac{Z_{n+1}}{Z_n} = \tau m \quad a.s.$$

The proof can be found in [GON 02].

Note that in theorem 4.6, $W \equiv 0$ on $\{Z_n \to 0\}$. Hence, we are interested in necessary and sufficient conditions for the sequence $\{W_n\}_{n\geq 0}$ to have a non-degenerate limit. In the remaining part of this chapter, we will present limit theorems for the three modes of convergence: almost sure, in L^1 and in L^2.

Before we continue, let us define the sequence $\{\tau(k)\}_{k\geq 0}$ by

$$\tau(k) := \tau - \frac{\varepsilon(k)}{k}, \qquad k = 1, 2, \ldots; \quad \tau(0) = 1,$$

assuming $\lim_{k\to\infty} \varepsilon(k)/k = \tau < \infty$. Obviously, if $\{\varepsilon(k)/k\}_{k\geq 1}$ converges monotonically to τ, then $\{|\tau(k)|\}_{k\geq 1}$ is non-increasing and converges to 0.

4.3.1. *Almost sure convergence*

We start with a necessary condition for W to be non-degenerate at 0.

THEOREM 4.8.– *Let* $\{Z_n\}_{n\geq 0}$ *be a CB process, such that*

$$\lim_{k\to\infty} \frac{\varepsilon(k)}{k} = \tau < \infty$$

and the assumptions of theorem 4.6 hold. If $P(W > 0) > 0$*, then on* $\{W > 0\}$

$$\sum_{k=0}^{\infty} |\tau(Z_k)| < \infty \qquad a.s.$$

The proof can be found in [GON 02].

As we pointed out in Chapter 1, in the theory of branching processes, the condition

$$E\left[X_{0,1} \log(1 + X_{0,1})\right] < \infty$$

is referred to as "$x \log x$ condition".

The next theorem (for its proof see [GON 02]) reveals that the "$x \log x$ condition" is sufficient for $P(W > 0) > 0$.

THEOREM 4.9.– *Let $\{Z_n\}_{n\geq 0}$ be a CB process, such that $\lim_{k\to\infty} \varepsilon(k)/k = \tau$, the assumptions of theorem 4.6 hold, and $\{\nu^2(k)/k\}_{k\geq 1}$ is bounded. Then*

$$P(W > 0) > 0$$

if and only if

$$E[X_{0,1} \log(1 + X_{0,1})] < \infty \quad \textit{and} \quad P\left(\sum_{k=0}^{\infty} |\tau(Z_k)| < \infty\right) > 0.$$

In the end, we present a more general result, which does not assume monotonic convergence of $\{\varepsilon(k)/k\}_{k\geq 1}$ (see [GON 06b] for the proof).

THEOREM 4.10.– *Let $\{Z_n\}_{n\geq 0}$ be a CB process. Suppose that:*

$$\textit{(i)} \qquad \lim_{k\to\infty} \frac{\varepsilon(k)}{k} = \tau < \infty, \qquad \tau m > 1.$$

(ii) There exists a positive and non-increasing function $\lambda(k)$ with $\sum_{k=1}^{\infty} \lambda(k)/k < \infty$, such that

$$|\tau(k)| \leq \lambda(k), \qquad k \geq 1.$$

(iii) There exists a positive and non-increasing function $\eta(x)$ with $\sum_{k=1}^{\infty} \eta(k)/k < \infty$, such that for k sufficiently large and $\alpha \in [1, 2]$

$$(\varepsilon(k)\eta(k))^{\alpha} \geq \max\left\{\frac{\varepsilon(k)}{m^{\alpha}} E[|X_{0,1} - m|^{\alpha}], E[|\phi_0(k) - \varepsilon(k)|^{\alpha}]\right\}.$$

Then $\lim_{n\to\infty} W_n = W$ a.s., where W is a finite and non-degenerate random variable, such that $P(W > 0) = P(Z_n \to \infty)$.

4.3.2. L^1–convergence

We will present without proofs necessary and sufficient conditions for $\{W_n\}_{n\geq 0}$ to converge in L^1 to a non degenerate at zero random variable.

Let us begin with a sufficient condition for the existence of the limit of $\{E[W_n]\}_{n\geq 0}$ as $n \to \infty$. The limiting random variable in theorem 4.11 is positive and finite provided that the process starts with a sufficiently large number of ancestors, so the non-extinction probability is positive. For the proof we refer to [GON 03].

THEOREM 4.11.– *Let $\{Z_n\}_{n\geq 0}$ be a CB process. Assume that:*

(i) $$\lim_{k\to\infty} \frac{\varepsilon(k)}{k} = \tau < \infty, \qquad \tau m > 1.$$

(ii) $\{|\tau(k)|\}_{k\geq 1}$ *is non-increasing and*

$$\sum_{k=1}^{\infty} \frac{|\tau(k)|}{k} < \infty.$$

Then:

1) for any $N \geq 1$ *such that* $q_N < 1, 0 < \lim_{n\to\infty} E[W_n \mid Z_0 = N] < \infty$;

2) there exists a finite random variable such that $\lim_{n\to\infty} W_n = W$ *a.s.*

COMMENT.– Clearly, assumption (ii) can be substituted by assumption (ii) of theorem 4.10. Moreover, since

$$\sum_{k=1}^{\infty} \frac{|\tau(k)|}{k} < \infty \qquad [4.21]$$

implies that the sequence $\{E[W_n]\}_{n\geq 0}$ is bounded, we obtain by theorem 4.11 that $W_n \to W$ a.s., provided $\{\varepsilon(k)/k\}_{k\geq 1}$ converges monotonically. Also, [4.21] and $\tau m > 1$ yield the a.s. convergence of $\{W_n\}_{n\geq 0}$ provided $\{\varepsilon(k)/k\}_{k\geq 1}$ converges to τ in an alternative way, such that $\{|\tau(k)|\}_{k\geq 1}$ is convergent and non-increasing, even if assumption (iii) of theorem 4.10 does not hold.

The following theorem gives a necessary condition for $\{W_n\}_{n\geq 0}$ to converge in L^1 to a non-degenerate at zero random variable (see [GON 03] for the proof).

THEOREM 4.12.– *Let $\{Z_n\}_{n\geq 0}$ be a CB process. Assume that*

$$\lim_{k\to\infty} \frac{\varepsilon(k)}{k} = \tau < \infty, \qquad \tau m > 1$$

and the convergence is monotonic. If $\lim_{n\to\infty} W_n = W$ *in* L^1, *where* W *is non-degenerate at 0, then* [4.21] *holds.*

Before presenting a sufficient condition for the L^1–convergence of $\{W_n\}_{n\geq 0}$ to a non-degenerate limit, we recall the notation $\ell_1(0) := 0$ and for $k = 1, 2, \ldots$

$$\ell_1(k) = E[|Z_{n+1} - \varepsilon(k)m| \mid Z_n = k].$$

Since $E[Z_{n+1} \mid Z_n = k] = \varepsilon(k)m$, we can interpret $\ell_1(k)$ as the mean absolute population size deviation (increment or decrement) given that the previous population size was k.

THEOREM 4.13.– *Let $\{Z_n\}_{n\geq 0}$ be a CB process, such that:*

(i) $\lim_{k\to\infty} \dfrac{\varepsilon(k)}{k} = \tau < \infty, \qquad \tau m > 1$ *and*

$\{|\tau(k)|\}_{k\geq 0}$ *is non–increasing.*

(ii) $\{\nu^2(k)/k\}_{k\geq 1}$ *is a bounded sequence.*

(iii) $\{\ell_1(k)/k\}_{k\geq 1}$ *is a non-increasing sequence.*

If the following series are convergent

$$\sum_{n=1}^{\infty} \frac{|\tau(n)|}{n} \quad \text{and} \quad \sum_{n=1}^{\infty} \frac{\ell_1(n)}{n^2},$$

then $\lim_{n\to\infty} W_n = W$ *in* L^1*, where* W *is finite and non-degenerate at 0.*

Due to its analytical complexity, the condition $\sum_{n=1}^{\infty} \ell_1(n)/n^2 < \infty$ can be difficult to check. In contrast, the assumptions in theorem 4.14 are easier to verify. Under these alternative assumptions, the "$x \log x$ condition" condition is necessary and sufficient for L^1 convergence to a non-degenerate limit.

THEOREM 4.14.– *Let $\{Z_n\}_{n\geq 0}$ be a CB process. Assume that:*

(i) $\lim_{k\to\infty} \dfrac{\varepsilon(k)}{k} = \tau < \infty, \qquad \tau m > 1$ *and*

$\{|\tau(k)|\}_{k\geq 0}$ *is non–increasing.*

(ii) $\{\nu^2(k)/k\}_{k\geq 1}$ *is bounded,*

(iii) $\displaystyle\sum_{k=1}^{\infty} \frac{|\tau(k)|}{k} < \infty.$

Then $\lim_{n\to\infty} W_n = W$ *in* L^1*, where* W *is finite and non-degenerate at 0 if and only if* $E[X_{0,1}\log(1+X_{0,1})] < \infty$.

The proofs of theorems 4.13 and 4.14 can be found in [GON 03]. Both results remain true if $\tau(n) = O(a_n)$ and $\ell_1(n) = O(b_n)$, where $\{a_n\}_{n\geq 0}$ and $\{b_n\}_{n\geq 0}$ satisfy the conditions for growth and summability imposed on the sequences $\{|\tau(n)|\}_{n\geq 0}$ and $\{\ell_1(n)\}_{n\geq 0}$, respectively. For instance, we can choose $\tau(n) = O(\log^{-\alpha} n)$ and $\ell_1(n) = O(\log^p n)$ for $\alpha > 1$ and $p > 0$.

4.3.3. L^2–convergence

In this last section, we will prove necessary and sufficient conditions for L^2–convergence of $\{W_n\}_{n\geq 0}$ to a non-degenerate random variable. First, we calculate the second-order moment of W_n.

PROPOSITION 4.1.– *Let* $\{Z_n\}_{n\geq 0}$ *be a CB process, such that*

$$\lim_{k\to\infty} \frac{\varepsilon(k)}{k} = \tau < \infty.$$

Then for $n = 0, 1, \ldots$

$$E[W_{n+1}^2] = Z_0^2 + \frac{1}{\tau^2}\sum_{k=0}^{n} E\left[W_k^2\left(\frac{\nu^2(Z_k)}{Z_k^2} - 2\tau\tau(Z_k) + \tau(Z_k)^2\right)\right] + \sigma^2 \sum_{k=0}^{n} \frac{E[\varepsilon(Z_k)]}{(\tau m)^{2(k+1)}}. \qquad [4.22]$$

Proof. As previously, denote $\mathcal{F}_n = \sigma(Z_0, \ldots, Z_n)$ for $n = 0, 1, \ldots$ We have

$$\begin{aligned}
E[Z_{n+1}^2] &= E[E[Z_{n+1}^2 \mid \mathcal{F}_n]] \\
&= E[\varepsilon(Z_n)\sigma^2 + m^2\nu^2(Z_n) + \varepsilon^2(Z_n)m^2] \\
&= E[\varepsilon(Z_n)\sigma^2 + m^2\nu^2(Z_n) + Z_n^2(\tau - \tau(Z_n))^2 m^2] \\
&= E\left[Z_n^2\tau^2 m^2 + Z_n^2 m^2\left(\frac{\nu^2(Z_n)}{Z_n^2} + \tau^2(Z_n) - 2\tau\tau(Z_n)\right)\right. \\
&\quad \left. + \sigma^2\varepsilon(Z_n)\right]
\end{aligned}$$

and therefore,

$$E[W_{n+1}^2] = E[W_n^2] + \frac{1}{\tau^2} E\left[W_n^2 \left(\frac{\nu^2(Z_n)}{Z_n^2} - 2\tau\tau(Z_n) + \tau^2(Z_n)\right)\right] + \sigma^2 \frac{E[\varepsilon(Z_n)]}{(\tau m)^{2(n+1)}}. \qquad [4.23]$$

Now, [4.22] follows from [4.23] upon iterations. ∎

PROPOSITION 4.2.– *Let $\{Z_n\}_{n\geq 0}$ be a CB process, such that:*

(i) $\lim_{k\to\infty} \frac{\varepsilon(k)}{k} = \tau < \infty, \qquad \tau m > 1$ *and*

$\{|\tau(k)|\}_{k\geq 0}$ *is non–increasing.*

(ii) $\displaystyle\sum_{k=1}^{\infty} \frac{|\tau(k)|}{k} < \infty.$

Then, for any $N \geq 1$ such that $q_N < 1$,

$$\sum_{n=0}^{\infty} \frac{E[\varepsilon(Z_n)]}{(\tau m)^{2(n+1)}} < \infty.$$

Proof. Since $\{\varepsilon(k)/k\}_{k\geq 1}$ is convergent, there exists $M > 0$ such that for any $k \geq 1$, $\varepsilon(k)/k \leq M$. Thus,

$$E[\varepsilon(Z_n)] \leq M, \qquad n = 0, 1, \ldots.$$

The assumptions *(i)* and *(ii)* imply $0 < \lim_{n\to\infty} E[W_n | Z_0 = N] < \infty$ (see theorem 4.11), and therefore there exists $M' > 0$ such that

$$E[W_n] \leq M', \qquad n = 0, 1, \ldots.$$

Therefore, since $\tau m > 1$, we obtain

$$\sum_{n=0}^{\infty} \frac{E[\varepsilon(Z_n)]}{(\tau m)^{2(n+1)}} \leq MM' \sum_{n=0}^{\infty} \frac{1}{(\tau m)^{n+2}} < \infty.$$

∎

We will prove a necessary condition for L^2–convergence of $\{W_n\}_{n\geq 0}$. Define

$$\xi(k) := \frac{\nu^2(k)}{k^2} - 2\tau\tau(k) + \tau^2(k), \qquad k = 1, 2, \ldots$$

THEOREM 4.15.– *Let $\{Z_n\}_{n\geq 0}$ be a CB process, such that*

$$\lim_{k\to\infty} \frac{\varepsilon(k)}{k} = \tau < \infty.$$

If $\lim_{n\to\infty} W_n = W$ in L^2, then

$$\inf_{k\geq 1} \left\{ \frac{\nu^2(k)}{k^2} - 2\tau\tau(k) + \tau^2(k) \right\} \leq 0.$$

Proof. Assume, on the contrary, there is $\alpha > 0$ such that $\xi(n) \geq \alpha$ for $n = 0, 1, \ldots$ It follows by [4.23] that for $n = 0, 1, \ldots$

$$\begin{aligned} E[W_{n+1}^2] &\geq E[W_n^2]\left(1 + \frac{\alpha}{\tau^2}\right) + \frac{\sigma^2 E[\varepsilon(Z_n)]}{(\tau m)^{2(n+1)}} \\ &\geq E[W_n^2]\left(1 + \frac{\alpha}{\tau^2}\right) \\ &\geq Z_0^2\left(1 + \frac{\alpha}{\tau^2}\right)^{n+1}. \end{aligned}$$

Hence, $\lim_{n\to\infty} E[W_n^2] = \infty$, which contradicts $\lim_{n\to\infty} W_n = W$ in L^2. This contradiction completes the proof. ∎

Next, we give a necessary condition for a non-degenerate L^2 limit.

THEOREM 4.16.– *Let $\{Z_n\}_{n\geq 0}$ be a CB process. Assume that:*

(i) $\lim_{k\to\infty} \dfrac{\varepsilon(k)}{k} = \tau < \infty, \qquad \tau m > 1$ *and*

$\{|\tau(k)|\}_{k\geq 0}$ *is non–increasing.*

(ii) there exists a positive function $\widetilde{\nu^2}(x)$ on $[0, \infty)$ such that $\widetilde{\nu^2}(k) = \nu^2(k)$, $k = 0, 1, \ldots$ and $\widetilde{\nu^2}(x)/x^2$ is non-increasing;

(iii) the terms of the sequence $\{\xi(k)\}_{k\geq 1}$ have the same sign.

If $\lim_{n\to\infty} W_n = W$ *in* L^2, *where* W *is non-degenerate, then*

$$\sum_{k=1}^{\infty} \frac{\nu^2(k)}{k^3} < \infty.$$

In the end, we will present sufficient conditions for L^2-convergence of $\{W_n\}_{n\geq 0}$ to a non-degenerate limit. For these two last results, we assume that $P(Z_n \to \infty) > 0$.

THEOREM 4.17.– *Let* $\{Z_n\}_{n\geq 0}$ *be a CB process, such that:*

$$\lim_{k\to\infty} \frac{\varepsilon(k)}{k} = \tau < \infty, \qquad \tau m > 1,$$

and $\{|\tau(k)|\}_{k\geq 1}$ *is non–increasing.*

Suppose there exist positive functions $\widetilde{\tau}(x)$ *and* $\widetilde{\nu}^2(x)$ *defined on* $[0, \infty)$, *where* $\widetilde{\tau}(x)$ *is non-increasing. Assume also that:*

(i) $\widetilde{\tau}(k) \geq |\tau(k)|$ *and* $\widetilde{\nu}^2(k) \geq \nu^2(k)$ *for* $k = 1, 2, \ldots$,

(ii) $g(x) := \widetilde{\nu}^2(x)x^{-2} + \widetilde{\tau}(x)^2 + 2\tau\widetilde{\tau}(x)$ *for* $x \in (0, \infty)$ *is non-increasing and both* $xg(x^{1/2})$ *and* $x\widetilde{\tau}^2(x^{1/2})$ *are concave on* $(0, \infty)$,

(iii) $$\sum_{k=1}^{\infty} \frac{\widetilde{\tau}(k)}{k} < \infty \quad \textit{and} \quad \sum_{k=1}^{\infty} \frac{\widetilde{\nu}^2(k)}{k^3} < \infty.$$

Then $\lim_{n\to\infty} W_n = W$ *in* L^2, *where* W *is finite and non-degenerate.*

COROLLARY 4.3.– *Let* $\{Z_n\}_{n\geq 0}$ *be a CB process such that*

$$\lim_{k\to\infty} \frac{\varepsilon(k)}{k} = \tau < \infty, \qquad \tau m > 1.$$

and $\{|\tau(k)|\}_{k\geq 1}$ *is non-increasing. Assume that:*

(i) $\{\nu^2(k)/k^2\}_{k\geq 1}$ *is non-increasing,*

(ii) the following two series are convergent

$$\sum_{k=1}^{\infty} \frac{|\tau(k)|}{k} \quad \textit{and} \quad \sum_{k=1}^{\infty} \frac{\nu^2(k)}{k^3}.$$

Then $\lim_{n\to\infty} W_n = W$ *in* L^2*, where* W *is finite and non-degenerate.*

The proofs of theorems 4.16 and 4.17 and corollary 4.3 can be found in [GON 05d]. A condition that implies a positive non-extinction probability, i.e. $P(Z_n \to \infty) > 0$ and makes the second series in (ii) convergent is $\nu^2(k) = O(k)$ as $k \to \infty$ (see theorem 3.2).

COMMENT.– L^α-convergence for $1 < \alpha \leq 2$ is studied in [NAK 94]. Note that in the case $\alpha = 2$, the assumption $\sum_{k=1}^{\infty} \nu^2(k)/k^3 < \infty$ in corollary 4.3 is weaker than the corresponding condition $\sum_{k=1}^{\infty} \nu^2(k)/k^2 < \infty$ in [NAK 94]. Also corollary 4.3 remains true if $\tau(n) = O(a_n)$ and $\nu^2(n) = O(b_n)$, where $\{a_n\}_{n\geq 0}$ and $\{b_n\}_{n\geq 0}$ are real numbers satisfying the conditions of growth and summability for $\{\tau(n)\}_{n\geq 0}$ and $\{\nu^2(n)\}_{n\geq 0}$, respectively. For example, we can choose $\tau(n) = O(\log^{-\alpha} n)$ and $\nu^2(n) = O(n^2 \log^{-\beta} n)$ for $\alpha, \beta > 1$.

4.4. Background and bibliographical notes

The first study on the asymptotic behavior of CB processes was done in [ZUB 74] considering deterministic control functions and supercritical processes. In particular, assuming control function $\phi(n) \sim \alpha n$ as $n \to \infty$ for a real positive number α, it was proved in [ZUB 74] that if $m\phi(n)/n$ converges fast enough to a constant greater than one, then the model has a geometric growth rate αm. In [BAG 86], under similar restrictions on the control function, an almost sure convergence theorem under the "$x \log x$ condition" was proved. A similar result was established in [MOL 98] for superadditive deterministic control functions.

The findings in [BAG 86] were extended to CB processes with random control functions in [NAK 90], following the lines of [KLE 89a]. New conditions for geometric growth in the supercritical case were given in [GON 02] and [GON 03], where almost sure and L^1-convergence were studied, applying arguments similar to those in [KLE 84] and [KLE 85] for population-size-dependent branching processes. L^α-convergence for $1 < \alpha \leq 2$ was first studied in [NAK 94]. Refinements in the case $\alpha = 2$ were obtained in [GON 05e]. Geometric growth, on the whole set where the process goes to infinity, was established in [GON 06b].

Subcritical CB processes were investigated in [GON 04b]. The results given in this paper were extended to MCB processes in [PUE 07]. The study of the critical case was developed in [GON 05d] and [GON 06c]. In [YAN 15], a survey on critical CB processes is provided, reviewing in particular the subclass of critical migration processes with different regimes of immigration and emigration.

Convergence of branching processes to a class of diffusion processes was first considered in [FEL 51] for GW processes and in [KAW 71] and [SRI 94] for GWI processes. More recent works on this subject are [KUE 11], [YAN 10], and [GON 12a].

Finally, it is important to point out that the results in the supercritical and critical cases were extended to the class of controlled multitype branching processes with population-size-dependent reproduction in [GON 05b], [GON 05c], [GON 06a] and [GON 07].

5

Statistics of CB Processes

"Prediction is very difficult, especially about the future".
– Niels Bohr

In this chapter, we gather and present statistical results for CB processes. The non-linear dynamics of the CB processes imposes serious challenges on the statistical analysis. We construct certain classes of estimators and study their properties from a theoretical point of view as well as their performance on simulations. Two asymptotic behaviors are of interest in the applications: when the time (or generation index) increases or when the number of ancestors is large. Although the latter is important, we deal exclusively with the former. We start with maximum likelihood estimators (MLEs) assuming that the complete family tree is observed (section 5.1). Expectation–maximization (EM) algorithms are developed to extend the constructed MLEs in case of incomplete data. In section 5.2, conditional weighted least squares estimators for the offspring mean are constructed. Their asymptotic behavior is investigated depending on the criticality of the processes. The issue of estimators' robustness is discussed in section 5.3 in the context of minimum disparity estimators (MDEs). Finally, in section 5.4, a Bayesian framework is discussed in case of complete data along with Markov chain Monte Carlo (MCMC) algorithm for the incomplete data situation.

5.1. Maximum likelihood estimation

In this section, we discuss MLEs of some parameters of interest in CB processes under various sample schemes. First, we consider data, which consist of the entire family tree up to the nth generation. In practice, observing the entire family tree is not common, so we proceed with MLEs based on observing either both generation sizes and number of progenitors or, alternatively, generation sizes only. We treat the data in the last two sampling schemes as incomplete and develop corresponding EM algorithms.

If the distribution of the control functions vary with the population size, then the problem of estimating its (control) parameters seems intractable unless the control process is assumed to have a stable structure over time. We will study CB processes with control functions having power series distribution. That is, for $k = 0, 1, \ldots$ and $n = 0, 1, \ldots$

$$P(\phi_n(k) = j) = \frac{a_k(j)\theta^j}{A(k,\theta)}, \qquad j = 0, 1, \ldots; \quad \theta \in \Theta, \qquad [5.1]$$

where $\Theta = \{\theta > 0 : 0 < A(k,\theta) < \infty$ for all $k = 0, 1, \ldots\}$, $A(k,\theta) = \sum_{j=0}^{\infty} a_k(j)\theta^j$, and $a_k(j) \geq 0$. Assume also the following regularity condition

$$A(k,\theta) = b(\theta)^{\varphi(k)}, \qquad \theta \in \Theta, \qquad k = 0, 1, \ldots, \qquad [5.2]$$

where $\varphi : \mathbf{N}_0 \to \mathbf{N}_0$ and $b(\cdot)$ is a differentiable function. Notice that, the control distribution depends on both θ, the so-called control parameter, and population size k. We can think of $\varphi(\cdot)$ as a function that provides deterministic control prior to the random one.

The mean of [5.1], denoted by $\varepsilon(k,\theta)$ or simply $\varepsilon(k)$, under condition [5.2], is given by

$$\begin{aligned}\varepsilon(k,\theta) &= \theta \frac{d}{d\theta} \log A(k,\theta) \\ &= \theta\varphi(k)\frac{d}{d\theta}\log b(\theta) \\ &= \varphi(k)\mu(\theta), \quad \text{say,}\end{aligned}$$

where

$$\mu := \mu(\theta) = \theta\frac{d}{d\theta}\log b(\theta).$$

The distribution [5.1] with [5.2] can be reparameterized, changing θ to μ. The new parameter μ can be interpreted as a migration parameter because of its role in the control. Indeed, given a population size and after applying a deterministic control given by the function $\varphi(\cdot)$, the model allows, on average, emigration, immigration or no migration depending on whether $\mu < 1$, $\mu > 1$ or $\mu = 1$, respectively.

The family of distributions [5.1] with [5.2] includes a lot of important discrete distributions. Here are some examples:

1) **Poisson distribution**. Let $\phi_n(k)$ follow a Poisson distribution with parameter $\theta\varphi(k)$. It is straightforward to see that [5.1] and [5.2] hold with

$$a_k(j) = \frac{\varphi^j(k)}{j!}, \qquad A(k,\theta) = e^{\theta\varphi(k)}.$$

2) **Binomial distribution**. Let $\phi_n(k)$ follow a Binomial distribution with parameters $\varphi(k)$ and q. Setting $\theta = q/(1-q)$, [5.1] and [5.2] hold with

$$a_k(j) = \binom{\varphi(k)}{j}, \qquad A(k,\theta) = (1+\theta)^{\varphi(k)}.$$

3) **Negative binomial distribution**. Let $\phi_n(k)$ follow a negative binomial distribution with parameters $\varphi(k)$ and q. In this case, setting $\theta = 1-q$, conditions [5.1] and [5.2] are verified with

$$a_k(j) = \binom{\varphi(k)+j-1}{j}, \qquad A(k,\theta) = \left(\frac{1}{1-\theta}\right)^{\varphi(k)}.$$

Finally, we will assume that the function $\varphi(\cdot)$ is known and, for some limiting results given later, asymptotically linear, i.e. there exists $\varphi > 0$, such that

$$\lim_{k\to\infty} \frac{\varphi(k)}{k} = \varphi. \qquad [5.3]$$

It is interesting to note that under the assumptions [5.1]–[5.3], for the mean growth rates $\tau_m(k)$ we have as $k \to \infty$

$$\begin{aligned}\tau_m(k) &= \frac{\varepsilon(k)}{k}m \\ &= \frac{\varphi(k)\mu}{k}m \\ &\to \varphi\mu m \\ &=: \tau_m \qquad \text{say,}\end{aligned}$$

where m is the offspring mean. We will denote τ_m **as criticality parameter**.

Our goal is, under the assumptions [5.1]–[5.3], to construct MLEs for both offspring distribution $p = \{p_k\}_{k\geq 0}$ and control parameter θ and their asymptotic properties. Consequently, we will estimate the offspring mean and variance as well as the migration parameter μ and the criticality parameter τ_m.

5.1.1. *MLE based on entire family tree up to* n*th generation*

Let us observe the entire family tree up to generation n (complete data). That is, the data consist of the values of the random variables

$$\{X_{l,i} : i = 1, 2, \ldots, \phi_l(Z_l); \quad l = 0, 1, \ldots, n-1\}. \qquad [5.4]$$

Let us introduce the random variables for $l = 0, 1, \ldots, n-1$ and $k = 0, 1, \ldots$

$$Z_l(k) := \sum_{i=1}^{\phi_l(Z_l)} \mathcal{I}\{X_{l,i} = k\},$$

Recall that $\mathcal{I}A$ is the indicator of set A. Intuitively, $Z_l(k)$ represents the number of individuals from generation l who have exactly k offspring. Clearly, for any l

$$\phi_l(Z_l) = \sum_{k=0}^{\infty} Z_l(k) \qquad \text{and} \qquad Z_{l+1} = \sum_{k=0}^{\infty} k Z_l(k).$$

Therefore, observing [5.4] leads to observing

$$\mathcal{Z}_n^* = \{Z_l(k) : l = 0, 1, \ldots, n-1; \quad k = 0, 1, \ldots\}. \qquad [5.5]$$

Denote by Δ_l for any $l \geq 0$, the total number of parents up to the lth generation, i.e.

$$\Delta_l := \sum_{j=0}^{l} \phi_j(Z_j).$$

Recall that $\sum_{j=0}^{l} Z_j$ equals the total number of individuals up to the lth generation and $\sum_{j=0}^{l} Z_j(k)$ is the total number of parents up to the lth generation, having exactly k offspring.

THEOREM 5.1.– *Let* $\{Z_n\}_{n\geq 0}$ *be a CB process with* [5.1]–[5.2]. *The MLEs of* p_k *for* $k = 0, 1, \ldots$ *and* θ*, based on* $\mathcal{Z}_n^*$*, are given by*

$$\widehat{p}_{k,n} = \frac{\sum_{j=0}^{n-1} Z_j(k)}{\Delta_{n-1}}, \quad k = 0, 1, \ldots, \quad \text{and} \quad \widehat{\theta}_n = \mu^{-1}\left(\frac{\Delta_{n-1}}{\sum_{l=0}^{n-1}\varphi(Z_l)}\right),$$

respectively, where $\mu^{-1}(\cdot)$ *denotes the inverse function of* $\mu(\cdot)$.

Proof. Referring to [5.5], for the likelihood function we have

$$\begin{aligned}\mathcal{L}(p,\theta \mid \mathcal{Z}_n^*) &= \theta^{\sum_{l=0}^{n-1}\phi_l^*}\prod_{l=0}^{n-1}\frac{1}{A(z_l,\theta)}\prod_{l=0}^{n-1}\frac{\phi_l^*!\, a_{z_l}(\phi_l^*)}{\prod_{k=0}^{\infty} z_l(k)!}\prod_{k=0}^{\infty}p_k^{z_l(k)} \\ &= \theta^{\sum_{l=0}^{n-1}\phi_l^*}\, b(\theta)^{-\sum_{l=0}^{n-1}\varphi(z_l)}\prod_{l=0}^{n-1}\frac{\phi_l^*!\, a_{z_l}(\phi_l^*)}{\prod_{k=0}^{\infty} z_l(k)!}\prod_{k=0}^{\infty}p_k^{z_l(k)},\end{aligned}$$

where $\phi_l^* := \sum_{k=0}^{\infty} z_l(k)$. Thus, the log-likelihood function can be written as

$$\ell(p,\theta \mid \mathcal{Z}_n^*) = f(p) + h(\theta) + K, \qquad [5.6]$$

where

$$f(p) := \sum_{l=0}^{n-1}\sum_{k=0}^{\infty} Z_l(k)\log p_k, \qquad h(\theta) := \Delta_{n-1}\log\theta - \sum_{l=0}^{n-1}\varphi(z_l)\log b(\theta)$$

and K does not depend on p and θ. To obtain the MLEs, we need to maximize $f(p) + h(\theta)$ subject to the constraints

$$\sum_{k=0}^{\infty} p_k = 1, \qquad p_k \geq 0.$$

Applying the method of Lagrange multipliers, we have the following equations for p to be an extreme point of $f(p)$

$$\frac{\partial}{\partial p_k}\left(f(p) + \lambda\left(1 - \sum_{k=0}^{\infty} p_k\right)\right) = 0, \qquad k - 0, 1, \ldots$$

$$\frac{\partial}{\partial \lambda}\left(f(p) + \lambda\left(1 - \sum_{k=0}^{\infty} p_k\right)\right) = 0.$$

The solution is given for $k = 0, 1, \ldots$ by

$$\begin{aligned}\widehat{p}_{k,n} &= \frac{\sum_{l=0}^{n-1} Z_l(k)}{\sum_{k=0}^{\infty}\sum_{l=0}^{n-1} Z_l(k)} \\ &= \frac{\sum_{l=0}^{n-1} Z_l(k)}{\Delta_{n-1}}.\end{aligned}$$

The solution $\{\widehat{p}_{k,n}\}_{k\geq 0}$ will result in maximum of $f(p)$ if $f(p)/\Delta_{n-1} = \sum_{k=0}^{\infty} \widehat{p}_{k,n} \log p_k$ has a maximum at $p_k = \widehat{p}_{k,n}$ for $k = 0, 1, \ldots$. Indeed, by Jensen's inequality,

$$\begin{aligned}\sum_{k=0}^{\infty} \widehat{p}_{k,n} \log p_k - \sum_{k=0}^{\infty} \widehat{p}_{k,n} \log \widehat{p}_{k,n} &= \sum_{k=0}^{\infty} \widehat{p}_{k,n} \log \frac{p_k}{\widehat{p}_{k,n}} \\ &\leq \log \left(\sum_{k=0}^{\infty} \widehat{p}_{k,n} \frac{p_k}{\widehat{p}_{k,n}} \right) \\ &= \log \sum_{k=0}^{\infty} p_k \\ &= 0.\end{aligned}$$

Hence $f(p)$ has a maximum at $p = \{\widehat{p}_{k,n}\}_{k\geq 0}$.

Furthermore, it can be shown that

$$\widehat{\theta}_n = \mu^{-1} \left(\frac{\Delta_{n-1}}{\sum_{l=0}^{n-1} \varphi(Z_l)} \right)$$

yields an extreme value of $h(\theta)$. The variance of [5.1], denoted by $\nu^2(k, \theta)$ or simply $\nu^2(k)$, is given by

$$\nu^2(k, \theta) = \theta \frac{d}{d\theta} \varepsilon(k, \theta).$$

Therefore, under the assumptions [5.1]–[5.2], we calculate the variance of the control functions to be

$$\nu^2(k, \theta) = \varphi(k)\theta\mu'(\theta).$$

Now, using $\varepsilon(k,\theta) = \varphi(k)\mu(\theta)$, we have

$$\begin{aligned}\left.\frac{d^2h(\theta)}{d\theta^2}\right|_{\theta=\widehat{\theta}_n} &= -\frac{1}{\theta^2}\left(\Delta_{n-1} + \sum_{l=0}^{n-1}\varphi(Z_l)(\mu'(\theta)\theta - \mu(\theta))\right)\Bigg|_{\theta=\widehat{\theta}_n} \\ &= -\frac{1}{\widehat{\theta}_n^2}\sum_{l=0}^{n-1}\nu^2(Z_l,\widehat{\theta}_n) \\ &< 0.\end{aligned}$$

Therefore, $\widehat{\theta}_n$ yields a maximum of $h(\theta)$ and $(\widehat{p}_n, \widehat{\theta}_n)$ maximizes $f(p) + h(\theta)$. ∎

Under the assumptions of theorem 5.1, it follows that the MLEs for m, σ^2 and μ are given by

$$\widehat{m}_n = \frac{\sum_{j=1}^n Z_j}{\Delta_{n-1}}, \quad \widehat{\sigma}_n^2 = \sum_{k=0}^{\infty}(k - \widehat{m}_n)^2\widehat{p}_{k,n}, \quad \widehat{\mu}_n = \frac{\Delta_{n-1}}{\sum_{l=0}^{n-1}\varphi(Z_l)},$$

respectively. Therefore, assuming [5.3], the criticality parameter τ_m is estimated by

$$\widehat{\tau}_{m,n} = \frac{\varphi\sum_{j=1}^n Z_j}{\sum_{l=0}^{n-1}\varphi(Z_l)}.$$

For simplicity, we will write $\widehat{p}_k$ and $\widehat{\tau}_m$ instead of $\widehat{p}_{k,n}$ and $\widehat{\tau}_{m,n}$.

The MLEs $\widehat{p}_k$ and $\widehat{m}_n$ are easily understood. $\widehat{p}_k$ estimates the probability p_k of having k offspring by the relative proportion of parents with k offspring. The offspring mean m is estimated by $\widehat{m}_n$, which equals the total number of offspring divided by the number of progenitors who have produced those offspring.

COMMENT.–

(i) Given the independence between the offspring and control variables, it is worth noting that to obtain the MLEs of the offspring distribution and its associated parameters, it is not necessary to impose any kind of parametric conditions on the control distribution. One can thus address this problem in a non-parametric framework, obtaining the same estimators for these three parameters.

(ii) Making use of lemma 2.13.2 in [JAG 75], it can be proved that $\widehat{m}_n$, $\widehat{\theta}_n$, $\widehat{\mu}_n$ and $\widehat{\tau}_m$ are also the MLEs of m, θ $\mu(\theta)$ and τ_m, respectively, based on the sample $\{Z_0, ..., Z_n, \phi_0(Z_0), ..., \phi_{n-1}(Z_{n-1})\}$. Moreover, $\widehat{\tau}_m$ is also the MLE of τ_m based on $\{Z_0, \dots, Z_n\}$ which can be shown using similar arguments.

We will investigate the **asymptotic properties** of the proposed estimators in the supercritical case. For the parameters of the offspring distribution, as was commented above, one does not need to assume that the control variables belong to a power series family of distributions. Instead, the only assumption needed is that $\{Z_n\}_{n\geq 0}$ satisfies the following conditions:

$$
\begin{array}{ll}
(i) & \lim_{k\to\infty} \frac{\varepsilon(k)}{k} = \tau < \infty \quad \text{and} \quad \sup_{k>0} \frac{\nu^2(k)}{k} =: M < \infty. \\
(ii) & \tau_m := \tau m > 1 \text{ and } Z_0 \text{ is large enough such that } P(Z_n \to \infty) > 0. \\
(iii) & \lim_{n\to\infty} \frac{Z_n}{\tau_m^n} = W < \infty \quad \text{a.s.,} \quad \text{such that } P(W > 0) > 0. \\
(iv) & \{W > 0\} = \{Z_n \to \infty\} \text{ a.s.}
\end{array}
\qquad [5.7]
$$

COMMENT.– In Chapters 3 and 4, sufficient conditions for (ii)–(iv) in [5.7] are provided. Moreover, on $\{Z_n \to \infty\}$, we have (see theorem 4.7)

$$
\lim_{n\to\infty} \frac{Z_{n+1}}{Z_n} = \tau_m \quad \text{a.s.} \qquad [5.8]
$$

We begin with a preliminary result needed to study the asymptotic properties of the proposed estimators.

PROPOSITION 5.1.– *Let $\{Z_n\}_{n\geq 0}$ be a CB process with* [5.7]. *Then almost surely on $\{Z_n \to \infty\}$:*

1) $$\lim_{n\to\infty} \frac{\phi_n(Z_n)}{Z_n} = \tau$$

2) $$\sum_{n=0}^{\infty} \frac{1}{\phi_n(Z_n)} < \infty$$

3) $$\lim_{n\to\infty} \frac{\sum_{j=0}^{n+1} Z_j}{\sum_{j=0}^{n} Z_j} = \tau_m$$

4) $$\lim_{n\to\infty} \frac{\Delta_n}{\sum_{j=0}^{n} Z_j} = \tau$$

5) $$\lim_{n\to\infty} \frac{\phi_n(Z_n)}{\Delta_n} = \frac{\tau_m - 1}{\tau_m}$$

6) $\lim_{n\to\infty} \frac{\phi_n(Z_n)}{\varepsilon(Z_n)} = 1$

Proof. Denote for $i = 1, \ldots, n$ and $n = 1, 2, \ldots$

$$\mathcal{G}_i := \sigma(Z_0, \ldots, Z_i, \phi_0(Z_0), \ldots, \phi_{i-1}(Z_{i-1}))$$

and let $(\Omega, \mathcal{A}, P)$ be a probability space on which all the variables are defined.

1) Define for $n \in \mathbf{N}^+$

$$X_n = \frac{\phi_n(Z_n)}{Z_n} \mathcal{I}\{Z_n > 0\}.$$

Therefore, for $\omega \in \Omega$

$$E[X_n \mid \mathcal{G}_n](\omega) = \sum_{k=1}^{\infty} \frac{\varepsilon(k)}{k} \mathcal{I}\{Z_n = k\}(\omega)$$

and

$$Var[X_n \mid \mathcal{G}_n](\omega) = \sum_{k=1}^{\infty} \frac{\nu^2(k)}{k^2} \mathcal{I}\{Z_n = k\}(\omega).$$

Let $\omega \in \{Z_n \to \infty\}$, such that $Z_n(\omega) = k_n > 0$ and $\liminf_{n\to\infty} k_{n+1}/k_n > 1$. Since $\sup_{k\geq 0} \nu^2(k)/k = M < \infty$, we have

$$\begin{aligned}
\sum_{n=0}^{\infty} Var[X_n \mid \mathcal{G}_n](\omega) &= \sum_{n=0}^{\infty} \frac{\nu^2(k_n)}{k_n^2} \\
&\leq M \sum_{n=0}^{\infty} \frac{1}{k_n} \\
&< \infty.
\end{aligned}$$

Moreover, since $Var[X_n|\mathcal{G}_n](\omega) = 0$ on $\omega \in \{Z_n \to 0\}$ and the dichotomy $P(Z_n \to 0) + P(Z_n \to \infty) = 1$ holds, we obtain

$$\sum_{n=0}^{\infty} Var[X_n|\mathcal{G}_n] < \infty \quad \text{a.s.}$$

Now, by Chebyshev's inequality

$$\sum_{n=0}^{\infty} P\left(|X_n - E[X_n|\mathcal{G}_n]| \geq \epsilon|\mathcal{G}_n\right) \leq \frac{1}{\epsilon^2} \sum_{n=0}^{\infty} Var[X_n|\mathcal{G}_n] < \infty \quad \text{a.s.}$$

Using the conditional Borel–Cantelli's lemma (see lemma A2.3, Appendix 2), we have

$$\left\{\sum_{n=0}^{\infty} P\left(|X_n - E[X_n|\mathcal{G}_n]| \geq \epsilon|\mathcal{G}_n\right) < \infty\right\}$$
$$= \bigcup_{k=0}^{\infty}\bigcap_{n=k}^{\infty}\{|X_n - E[X_n|\mathcal{G}_n]| < \epsilon\} \qquad \text{a.s.,}$$

hence $\lim_{n\to\infty} X_n - E[X_n|\mathcal{G}_n] = 0$ a.s. on $\{Z_n \to \infty\}$. This, in turn, implies

$$\lim_{n\to\infty}\frac{\phi_n(Z_n)}{Z_n} = \lim_{k\to\infty}\frac{\varepsilon(k)}{k}$$
$$= \tau \qquad \text{a.s. on } \{Z_n \to \infty\}.$$

2) It follows from *1)* and [5.8].

3) Setting $Y_n := \sum_{j=0}^{n} Z_j$, we write

$$\frac{Y_{n+1}}{Y_n} = \frac{Z_0 + Z_1}{Y_n} + \frac{1}{Y_n}\sum_{k=1}^{n}\frac{Y_k - Y_{k-1}}{Z_k}Z_{k+1}$$

Since $Y_n \to \infty$ on $\{Z_n \to \infty\}$, applying Cèsaro's lemma (see lemma A2.2, Appendix 2) we obtain

$$\lim_{n\to\infty}\frac{Y_{n+1}}{Y_n} = \lim_{n\to\infty}\frac{Z_0 + Z_1}{Y_n} + \lim_{n\to\infty}\frac{1}{Y_n}\sum_{k=1}^{n}(Y_k - Y_{k-1})\frac{Z_{k+1}}{Z_k}$$
$$= \tau_m \qquad \text{a.s. on } \{Z_n \to \infty\}.$$

4) Similarly to *3)*, taking into account 1) and applying Cèsaro's lemma.

5) It follows from *1)* and [5.8] that

$$\lim_{n\to\infty}\frac{\phi_n(Z_n)}{\phi_{n+1}(Z_{n+1})} = \frac{1}{\tau_m} < 1 \qquad \text{a.s. on } \{Z_n \to \infty\}.$$

Thus, for almost every $\omega \in \{Z_n \to \infty\}$, there exists $n_0 = n_0(\omega)$ such that $\phi_{n+1}(Z_{n+1}(\omega)) \neq \phi_n(Z_n(\omega))$ for all $n \geq n_0$. Therefore,

$$\lim_{n\to\infty} \frac{\phi_n(Z_n(\omega))}{\phi_n(Z_n(\omega)) - \phi_{n-1}(Z_{n-1}(\omega))} = \lim_{n\to\infty} \frac{1}{1 - \dfrac{\phi_{n-1}(Z_{n-1}(\omega))}{\phi_n(Z_n(\omega))}}$$

$$= \frac{\tau_m}{\tau_m - 1}.$$

Now, since $\Delta_n(\omega) = \sum_{j=0}^{n} \phi_j(Z_j(\omega))$, we have for $n \geq n_0$

$$\frac{\Delta_n(\omega)}{\phi_n(Z_n(\omega))} = \frac{1}{\phi_n(Z_n(\omega))} \sum_{j=0}^{n} \phi_j(Z_j(\omega))$$

$$= \frac{1}{\phi_n(Z_n(\omega))} \left[\sum_{k=0}^{n_0} \phi_k(Z_k(\omega)) + \sum_{k=n_0+1}^{n} \frac{\phi_k(Z_k(\omega)) - \phi_{k-1}(Z_{k-1}(\omega))}{1 - \frac{\phi_{k-1}(Z_{k-1}(\omega))}{\phi_k(Z_k(\omega))}} \right]$$

$$\to \frac{\tau_m}{\tau_m - 1} \quad \text{a.s. on } \{Z_n \to \infty\},$$

because, by Cèsaro's lemma, the term in the first sum converges to 0 and the term in the second sum converges to $(\tau_m - 1)/\tau_m$ on $\{Z_n \to \infty\}$.

6) It is a direct consequence of *1)*.

Proposition 5.1 is proved. ∎

Next, we study the asymptotic properties of the MLEs $\widehat{p}_k$, $\widehat{m}_n$ and $\widehat{\sigma}_n^2$. For simplicity, denote $\mathcal{D} := \{Z_n \to \infty\}$ and $P_{\mathcal{D}}(\cdot) := P(\cdot \mid \mathcal{D})$. Note that, the following result is true whether or not conditions [5.1] and [5.2] hold.

THEOREM 5.2.– *Let $\{Z_n\}_{n\geq 0}$ be a CB process. Under the assumptions* [5.7], *the following statements are true.*

1) $\widehat{p}_k$, $\widehat{m}_n$ and $\widehat{\sigma}_n^2$ are strongly consistent on $\{Z_n \to \infty\}$.

2) Let $\Phi(\cdot)$ denote the c.d.f. of the standard Normal distribution. If P' is a probability measure dominated by $P_{\mathcal{D}}$, then for any real x

(2.1) $$\lim_{n\to\infty} P'\left(\sqrt{\frac{\Delta_{n-1}}{p_k(1-p_k)}}\,(\widehat{p}_k - p_k) \le x\right) = \Phi(x),$$

(2.2) $$\lim_{n\to\infty} P'\left(\sqrt{\frac{\Delta_{n-1}}{\sigma^2}}\,(\widehat{m}_n - m) \le x\right) = \Phi(x),$$

(2.3) if, in addition, $E[X_{0,1}^4] < \infty$, then

$$\lim_{n\to\infty} P'\left(\sqrt{\frac{\Delta_{n-1}}{Var[(X_{0,1}-m)^2]}}\,(\widehat{\sigma}_n^2 - \sigma^2) \le x\right) = \Phi(x).$$

COMMENT.– Theorem 5.1 and lemma 2.3 in [GUT 91] yield that *2)* also holds for $P(\cdot|Z_n > 0)$. Thus, making use of theorem 5.2, Slutsky's theorem (see Appendix 2), and assuming $Z_n > 0$, we can construct asymptotic confidence intervals for the parameters p, m and σ^2. Namely, the asymptotic confidence intervals for p_k and m at the $1-\alpha$ level are given by

$$\widehat{p}_k \pm z_\alpha\sqrt{\frac{\widehat{p}_k(1-\widehat{p}_k)}{\Delta_{n-1}}} \quad \text{and} \quad \widehat{m}_n \pm z_\alpha\sqrt{\frac{\widehat{\sigma}_n^2}{\Delta_{n-1}}},$$

respectively, where $1-\Phi(z_\alpha) = \alpha/2$. In case of estimating offspring variance, to avoid the dependence on the unknown parameter $Var[(X_{0,1}-m)^2]$, it can be substituted by a consistent estimator (as follows by Slutsky's theorem). In fact, a strongly consistent estimator for $Var[(X_{0,1}-m)^2]$ is the MLE estimator

$$\sum_{k=0}^{\infty}(k-\widehat{m}_n)^4\widehat{p}_k - \widehat{\sigma}_n^4.$$

Proof of theorem 5.2. 1) Applying the strong law of large numbers for martingales (see theorem A1.3 in Appendix 1), we will prove that $\widehat{p}_k$ is strongly consistent, i.e. for a fixed $k \ge 0$

$$\begin{aligned}\lim_{n\to\infty}\widehat{p}_k &= \lim_{n\to\infty}\frac{1}{\Delta_{n-1}}\sum_{i=1}^{n}\sum_{j=1}^{\phi_{i-1}(Z_{i-1})}\mathcal{I}\{X_{i-1,j}=k\} \\ &= p_k \quad \text{a.s. on } \{Z_n\to\infty\}.\end{aligned} \qquad [5.9]$$

For simplicity, consider $P(Z_n \to \infty) = 1$. Denote, for $i = 1, 2, \ldots$,

$$\mathcal{H}_i := \sigma(X_{l-1,j}, \phi_{l-1}(k) : j \geq 1, k \geq 0, 1 \leq l \leq i).$$

Also, denote

$$V_i(k) := \sum_{j=1}^{\phi_{i-1}(Z_{i-1})} (\mathcal{I}\{X_{i-1,j} = k\} - p_k).$$

It can be seen that $\{V_i(k), \mathcal{H}_i\}_{i\geq 1}$ is a martingale difference. Using the new notation, we write

$$\widehat{p}_k - p_k = \frac{1}{\Delta_{n-1}} \sum_{i=1}^{n} V_i(k).$$

Recalling that $Y_n = \sum_{j=0}^{n} Z_j$ and taking into account proposition 5.1, part 4), we see that to prove [5.9] it is sufficient to show

$$\lim_{n\to\infty} \frac{1}{Y_{n-1}} \sum_{i=1}^{n} V_i(k) = 0 \quad \text{a.s.} \qquad [5.10]$$

Since $Y_{n-1} \to \infty$, in order to apply theorem A1.3 from Appendix 1, we need to verify that

$$\sum_{i=1}^{\infty} \frac{E[V_i^2(k) \mid \mathcal{H}_{i-1}]}{Y_{i-1}^2} < \infty \text{ a.s.}$$

Let $M := \sup_{n\geq 0} \Delta_n / Y_n$ and $N := \sup_{n\geq 0} \varepsilon(Z_n)/\phi_n(Z_n)$. Both M and N are finite a.s. by proposition 5.1, parts 4) and 6), respectively. Therefore,

$$\sum_{i=1}^{\infty} \frac{E[V_i^2(k) \mid \mathcal{H}_{i-1}]}{Y_{i-1}^2}$$

$$= \sum_{i=1}^{\infty} \frac{E\left[Var\left[\sum_{j=1}^{\phi_{i-1}(Z_{i-1})} \mathcal{I}\{X_{i-1,j} = k\} \middle| \phi_{i-1}(Z_{i-1})\right]\right]}{Y_{i-1}^2}$$

$$= \sum_{i=1}^{\infty} \frac{\varepsilon(Z_{i-1}) p_k (1 - p_k)}{Y_{i-1}^2}$$

$$= p_k(1 - p_k) \sum_{i=1}^{\infty} \frac{\varepsilon(Z_{i-1})}{\phi_{i-1}(Z_{i-1})} \cdot \frac{\phi_{i-1}(Z_{i-1})}{\Delta_{i-1}^2} \cdot \frac{\Delta_{i-1}^2}{Y_{i-1}^2}$$

$$\leq p_k(1-p_k)NM^2 \sum_{i=1}^{\infty} \frac{1}{\phi_{i-1}(Z_{i-1})} < \infty \quad \text{a.s.},$$

the last inequality is due to $\phi_{i-1}(Z_{i-1}) \leq \Delta_{i-1}$ and proposition 5.1, part 2).

The strong consistency of $\widehat{m}_n$ follows from proposition 5.1, parts 2)–5). Since $\widehat{m}_n$ and $\widehat{p}_k$ are strongly consistent on $\{Z_n \to \infty\}$, we have that $\widehat{\sigma}_n^2$ is strongly consistent too.

2) The key argument in proving (2.1) and (2.2) is to rewrite

$$\widehat{p}_k - p_k \stackrel{d}{=} \frac{1}{\Delta_{n-1}} \sum_{l=1}^{\Delta_{n-1}} (\mathcal{I}\{X_l = k\} - p_k),$$

$$\widehat{m}_n - m \stackrel{d}{=} \frac{1}{\Delta_{n-1}} \sum_{l=1}^{\Delta_{n-1}} (X_l - m),$$

where recall that $\stackrel{d}{=}$ stands for equality in distribution, and $\{X_l\}_{l\geq 1}$ is a sequence of i.i.d. random variables with common distribution the same as the offspring distribution. The results are obtained by applying a functional limit theorem (see theorem A2.3 in Appendix 2) as it was done, *mutatis mutandis*, in the proofs of theorems 3.2 and 4.2 in [GON 04a] for CB processes with deterministic control function. In particular, to prove (2.1) we apply theorem A2.3 in Appendix 2 with $\xi_n = \mathcal{I}\{X_n = k\} - p_k$, $a_n = (\tau_m^{n+1} - 1)/(\tau_m - 1)$, $\nu_n = \Delta_n$, $\widetilde{W} = \tau W$, and for $0 \leq t \leq 1$,

$$Y_{n-1}(t,\omega) = \frac{1}{\sqrt{p_k(1-p_k)\Delta_{n-1}(\omega)}} \sum_{l=1}^{[\Delta_{n-1}(\omega)t]} (\mathcal{I}\{X_l = k\}(\omega) - p_k),$$

where recall that $[x]$ denotes the integer part of x.

Since $\tau_m > 1$, we have $\lim_{n\to\infty} a_n = \lim_{n\to\infty}(\tau_m^{n+1} - 1)/(\tau_m - 1) = \infty$. Moreover,

$$\begin{aligned}
\frac{\nu_n}{a_n} &= \Delta_n \frac{\tau_m - 1}{\tau_m^{n+1} - 1} \\
&= \frac{1}{\sum_{j=0}^{n} \tau_m^j} \sum_{i=0}^{n} \tau_m^i \frac{\phi_i(Z_i)}{\tau_m^i}.
\end{aligned}$$

Note that [5.7], part (iii) and proposition 5.1, part 1) imply $\lim_{n\to\infty} \phi_n(Z_n)/\tau_m^n = \tau W$ a.s. on $\{Z_n \to \infty\}$. Therefore, by Cèsaro's lemma, we have that $\lim_{n\to\infty} \nu_n/a_n = \tau W$ a.s. on $\{Z_n \to \infty\}$. Thus, the hypotheses of theorem A2.3 in Appendix 2 hold, and consequently, for any real number x

$$\lim_{n\to\infty} P'\left(\sqrt{\frac{\Delta_{n-1}}{p_k(1-p_k)}}(\widehat{p}_k - p_k) \leq x\right) = \Phi(x).$$

The proof of (2.2) is similar.

Finally, to prove (2.3), we adapt the proof of theorem 3.1 in [GON 05a] for CB processes with deterministic control function. First, the result can be obtained for $\sum_{k=0}^{\infty}(k-m)^2\widehat{p}_k$, i.e. for known m. In this situation since

$$\sum_{k=0}^{\infty}(k-m)^2\widehat{p}_k \stackrel{d}{=} \frac{1}{\Delta_{n-1}}\sum_{l=1}^{\Delta_{n-1}}(X_l - m)^2,$$

the result follows by applying again theorem A2.3 from Appendix 2 setting $\xi_n = X_n - m$ and a_n, ν_n, and $\widetilde{W}$ as above, and for $0 \leq t \leq 1$

$$Y_{n-1}(t,\omega) = \sqrt{\frac{\Delta_{n-1}(\omega)}{Var[(X_{0,1}-m)^2]}}\sum_{j=1}^{[\Delta_{n-1}(\omega)t]}\left((X_j(\omega)-m)^2 - \sigma^2\right).$$

Now, notice that

$$\sum_{k=0}^{\infty}(k-m)^2\widehat{p}_k - \widehat{\sigma}_n^2 = (\widehat{m}_n - m)^2.$$

Therefore, from (2.2) and $\sigma^2/\sqrt{\Delta_{n-1}} \xrightarrow{P'} 0$ we have

$$\begin{aligned}\sqrt{\Delta_{n-1}}\left(\sum_{k=0}^{\infty}(k-m)^2\widehat{p}_k - \widehat{\sigma}_n^2\right) &= \sqrt{\Delta_{n-1}}(\widehat{m}_n - m)^2 \\ &= \frac{\Delta_{n-1}(\widehat{m}_n - m)^2}{\sigma^2}\frac{\sigma^2}{\sqrt{\Delta_{n-1}}} \\ &\xrightarrow{d} 0 \quad \text{as} \quad n \to \infty\end{aligned}$$

by Slutsky's theorem and therefore

$$\sqrt{\Delta_{n-1}}\left(\sum_{k=0}^{\infty}(k-m)^2\widehat{p}_k - \widehat{\sigma}_n^2\right) \xrightarrow{P'} 0 \qquad \text{as} \quad n\to\infty.$$

The claim in (2.3) follows by Slutsky's theorem and the fact that it is true when m is known. ∎

Consider now **MLEs for the parameters of the control law**. Recall that if the control functions belong to the power series family of distributions, then [5.7](i) holds trivially, with $\tau = \varphi\mu$, and hence $\tau_m = m\varphi\mu$. The following result is true.

THEOREM 5.3.– *Let $\{Z_n\}_{n\geq 0}$ be a CB process with* [5.1]–[5.3] *and* [5.7]. *The following statements then hold.*

1) $\widehat{\theta}_n$, $\widehat{\mu}_n$ and $\widehat{\tau}_m$ are strongly consistent on $\{Z_n\to\infty\}$.

2) Let $\Phi(\cdot)$ denote the standard Normal c.d.f. If, for each $l\geq 0$ and $z\geq 0$,

$$\phi_l(z) \stackrel{d}{=} \sum_{s=1}^{\varphi(z)} X_s(l,z),$$

where $\{X_s(l,z) : 1\leq s\leq \varphi(z),\ z\geq 0,\ l\geq 0\}$ are i.i.d. random variables with mean $\mu(\theta)$ and variance $\theta\mu'(\theta)$, then for any real number x

$$(2.1)\ \lim_{n\to\infty} P_{\mathcal{D}}\left(\sqrt{\frac{\sum_{i=0}^{n-1}\varphi(Z_i)}{\theta\mu'(\theta)}}\,(\widehat{\mu}_n - \mu(\theta)) \leq x\right) = \Phi(x),$$

$$(2.2)\ \lim_{n\to\infty} P_{\mathcal{D}}\left(\sqrt{\frac{\sum_{i=0}^{n-1}\varphi(Z_i)}{\varphi^2(\sigma^2\mu(\theta)+m^2\theta\mu'(\theta))}}\,(\widehat{\tau}_m - \tau_m) \leq x\right) = \Phi(x).$$

COMMENT.– (i) Condition set out in theorem 5.3, part 2) is satisfied by the control distributions introduced previously as examples.

(ii) Theorem 5.3, part 2) also holds for $P(\cdot|Z_n>0)$. Again, assuming $Z_n>0$, from this theorem and Slutsky's theorem, and replacing the values m, σ^2, θ

and $\mu'(\theta)$ by $\widehat{m}_n$, $\widehat{\sigma}_n^2$, $\widehat{\theta}_n$ and $\mu'(\widehat{\theta}_n)$, respectively, one can obtain asymptotic confidence intervals for the parameters $\mu(\theta)$ and τ_m at $1-\alpha$ level

$$\widehat{\mu}_n \pm z_\alpha \sqrt{\frac{\widehat{\theta}_n \mu'(\widehat{\theta}_n)}{\sum_{i=0}^{n-1}\varphi(Z_i)}} \quad \text{and} \quad \widehat{\tau}_m \pm z_\alpha \sqrt{\frac{\varphi^2(\widehat{\sigma}_n^2 \mu(\widehat{\theta}_n) + \widehat{m}_n^2 \theta \mu'(\widehat{\theta}_n))}{\sum_{i=0}^{n-1}\varphi(Z_i)}},$$

where $1-\Phi(z_\alpha) = \alpha/2$.

(iii) $\widehat{\tau}_m$ is strongly consistent on $\{Z_n \to \infty\}$ for CB processes with [5.7] only.

Proof of theorem 5.3. 1) It follows directly from proposition 5.1, part 4).

Proof of (2.1). For simplicity, we suppose $P(Z_n \to \infty) = 1$. Denote for $i = 1, \dots, n$ and $n = 1, 2, \dots$

$$D_i := \phi_{i-1}(Z_{i-1}) - \mu(\theta)\varphi(Z_{i-1}) \quad \text{and} \quad H_n := \sum_{i=0}^{n-1} \varphi(Z_i).$$

We have

$$\sqrt{H_n}\,(\widehat{\mu} - \mu(\theta)) \qquad [5.11]$$

$$= \frac{1}{\sqrt{H_n}} \sum_{i=1}^{n} D_i \left(\frac{\sqrt{\varphi(Z_{i-1})+1} - \sqrt{\tau_m^{i-1}\varphi W}}{\sqrt{\varphi(Z_{i-1})+1}} + \sqrt{\varphi W}\sqrt{\frac{\tau_m^{i-1}}{\varphi(Z_{i-1})+1}} \right),$$

where W is introduced in [5.7](iii).

Denote by $o_P(\cdot)$ the stochastic order analogue of $o(\cdot)$, that is $X_n = o_P(Y_n)$ if and only if for any $\epsilon > 0$

$$\lim_{n\to\infty} P(|X_n| \geq \epsilon |Y_n|) = 0.$$

Taking into account that

$$\lim_{n\to\infty} \frac{1}{\tau_m^n} H_n = \frac{\varphi W}{\tau_m - 1} \qquad \text{a.s.,}$$

it follows that to show *(2.1)* it is sufficient to prove, as $n \to \infty$,

$$(I) := \sum_{i=1}^{n} D_i \left(\frac{\sqrt{\varphi(Z_{i-1})+1} - \sqrt{\tau_m^{i-1}\varphi W}}{\sqrt{\varphi(Z_{i-1})+1}} \right) = o_P\left(\sqrt{\tau_m^n}\right) \qquad [5.12]$$

and

$$(II) := \sqrt{\tau_m - 1} \sum_{i=1}^{n} \frac{D_i}{\sqrt{\tau_m^{n-i+1}(\varphi(Z_{i-1}) + 1)}} \xrightarrow{d} N(0, \theta\mu'(\theta)). \quad [5.13]$$

For each $n \geq 0$ set

$$C_{1,n} := \sum_{i=1}^{n} \sqrt{\tau_m^{i-1}} \left(\sqrt{\frac{\varphi(Z_{i-1}) + 1}{\tau_m^{i-1}}} - \sqrt{\varphi W} \right)^2,$$

$$C_{2,n} := \sum_{i=1}^{n} \frac{D_i^2 \sqrt{\tau_m^{i-1}}}{\varphi(Z_{i-1}) + 1}.$$

Then, applying Cauchy-Schwarz inequality, $|(I)| \leq \sqrt{C_{1,n} C_{2,n}}$. By [5.7](iii)

$$\lim_{i \to \infty} \sqrt{\frac{\varphi(Z_{i-1}) + 1}{\tau_m^{i-1}}} - \sqrt{\varphi W} = 0 \quad \text{a.s.}$$

and consequently, using Cèsaro's lemma,

$$C_{1,n} = o_P \left(\sum_{i=1}^{n} \sqrt{\tau_m^{i-1}} \right) = o_P \left(\sqrt{\tau_m^n} \right).$$

Now, setting $\mathcal{G}_i := \sigma(Z_0, \ldots, Z_i, \phi_0(Z_0), \ldots, \phi_{i-1}(Z_{i-1}))$ for $i = 1, 2, \ldots$ and $\mathcal{G}_0 := \sigma(Z_0)$ and using

$$E[D_i^2 | \mathcal{G}_{i-1}] = \theta\varphi(Z_{i-1})\mu'(\theta), \quad i \geq 1, \quad [5.14]$$

we obtain as $n \to \infty$

$$E[C_{2,n}] = O \left(\sum_{i=1}^{n} \sqrt{\tau_m^{i-1}} \right) = O \left(\sqrt{\tau_m^n} \right),$$

which, in turn, implies as $n \to \infty$

$$|C_{2,n}| = O_P \left(\sum_{i=1}^{n} \sqrt{\tau_m^{i-1}} \right) = O_P \left(\sqrt{\tau_m^n} \right),$$

where $O_P(\cdot)$ denotes the stochastic order analogue of $O(\cdot)$, i.e. $X_n = O_P(Y_n)$ if and only if there exists a real number M, such that

$$\lim_{n \to \infty} P(|X_n| \geq M|Y_n|) = 0.$$

Hence, [5.12] follows.

To prove [5.13], let $\gamma_{nj} := D_{n-j+1}/\sqrt{\varphi(Z_{n-j})+1}$ for $j = 1, \dots, n$. Then

$$\sqrt{\tau_m - 1} \sum_{i=1}^{n} \frac{1}{\sqrt{\tau_m^{n-i+1}}} \frac{D_i}{\sqrt{\varphi(Z_{i-1})+1}} \tag{5.15}$$

$$= \sqrt{\tau_m - 1} \sum_{j=1}^{n} \frac{1}{\sqrt{\tau_m^j}} \frac{D_{n-j+1}}{\sqrt{\varphi(Z_{n-j})+1}}$$

$$= \sqrt{\tau_m - 1} \left(\sum_{j=1}^{J} \frac{\gamma_{nj}}{\sqrt{\tau_m^j}} + \sum_{j=J+1}^{n} \frac{\gamma_{nj}}{\sqrt{\tau_m^j}} \right)$$

$$=: U_{1,J}(n) + U_{J+1,n}(n), \quad \text{say.}$$

For $J \geq 1$ and given $(t_1, \dots, t_J)$, it can be proved, using the same arguments as in the proof of theorem 1 in [HEY 71] that under the assumption in (2)

$$\lim_{n\to\infty} E\left[\exp\left\{ i \sum_{j=1}^{J} t_j \frac{\gamma_{nj}}{\sqrt{\tau_m^j}} \right\}\right] = \exp\left\{ -\frac{1}{2}\theta\mu'(\theta) \sum_{j=1}^{J} \frac{t_j^2}{\tau_m^j} \right\}.$$

Hence, the vector $(\gamma_{n1}/\sqrt{\tau_m}, \dots, \gamma_{nJ}/\sqrt{\tau_m^J})$, for each $J \geq 1$, is asymptotically multivariate normal as $n \to \infty$, and therefore

$$U_{1,J}(n) \xrightarrow{d} N\left(0,\ \theta\mu'(\theta)(\tau_m - 1) \sum_{j=1}^{J} \frac{1}{\tau_m^j}\right) \quad \text{as} \quad n \to \infty \tag{5.16}$$

$$\xrightarrow{d} N\left(0,\ \theta\mu'(\theta)\right) \quad \text{as} \quad J \to \infty.$$

Now, from Chebyschev's inequality, [5.14], and [5.15] it follows that for every $n \geq 0$ and $\epsilon > 0$

$$P\left(|U_{J+1,n}(n)| > \epsilon\right) \leq \frac{(\tau_m - 1)\theta\mu'(\theta)}{\epsilon^2} \sum_{j=J+1}^{\infty} \frac{1}{\tau_m^j}.$$

Thus, for some constant k_0 as $J \to \infty$ we have

$$\limsup_{n\to\infty} P\left(|U_{J+1,n}(n)| > \epsilon\right) \leq k_0 \sum_{j=J+1}^{\infty} \frac{1}{\tau_m^j} \to 0. \tag{5.17}$$

Now, [5.13] follows from [5.15]–[5.17] and referring to theorem 25.5 in [BIL 95].

The claim in (2.2) can be proved using the same arguments as in the proof of (2.1), setting for $i = 1, \ldots, n$ and $n = 1, 2, \ldots$

$$D_i = \varphi Z_i - \tau_m \varphi(Z_{i-1}) \qquad \text{and} \qquad \mathcal{F}_i = \sigma(Z_0, \ldots, Z_i).$$

Consequently, for $i = 1, 2, \ldots$ we obtain

$$E[D_i^2 | \mathcal{F}_{i-1}] = \varphi^2(\sigma^2 \mu(\theta) + m^2 \theta \mu'(\theta))\varphi(Z_{i-1})$$

and therefore,

$$\lim_{n\to\infty} E\left[\exp\left\{i \sum_{j=1}^{J} \frac{t_j \gamma_{nj}}{\sqrt{\tau_m^j}}\right\}\right] = \exp\left\{-\frac{1}{2}\varphi^2(\sigma^2\mu(\theta) + m^2\theta\mu'(\theta)) \sum_{j=1}^{J} \frac{t_j^2}{\tau_m^j}\right\}.$$

The proof of the theorem is complete. ∎

5.1.2. *EM algorithms for incomplete data*

In section 5.1.1, we obtained MLEs for the parameters of interest based on the complete sample $\mathcal{Z}_n^*$. In practice, however, it might be difficult to observe the entire family tree. A more realistic assumption on the collected data would be that only the total number of individuals and progenitors in each generation can be observable. We approach the problem of the MLE under the reduced sample as an incomplete data estimation problem. Making use of the EM algorithm, we treat the unobservable part of $\mathcal{Z}_n^*$ as *hidden* variables. Starting with an initial offspring distribution $\{p_k^{(0)}\}_{k\geq 0}$ and an initial value of the control parameter $\theta^{(0)}$, we construct a sequence $\{(p^{(i)}, \theta^{(i)})\}_{i\geq 0}$ that converges to the MLE for (p, θ). This iterative method has two steps: E and M, which are repeated until a convergence criterion is met. In the E step, we calculate the expectation of the log-likelihood using the distribution of the unobserved data. In the following M step, we find the values of the parameters that maximize this expectation.

We will construct the MLEs for the main parameters of the model assuming that only the set of random variables

$$\bar{\mathcal{Z}}_n = \{Z_0, \ \ldots, Z_n, \ \phi_0(Z_0), \ldots, \phi_{n-1}(Z_{n-1})\}$$

are observed. As we pointed out in section 5.1.1, the constructed estimators $\widehat{m}_n$, $\widehat{\theta}_n$ and $\widehat{\theta}_n$ are still MLEs for m, θ and $\mu(\theta)$, respectively, given the reduced data $\bar{\mathcal{Z}}_n$. Hence, it only remains to construct MLEs for p and σ^2 based on $\bar{\mathcal{Z}}_n$, although below we will discuss the general method for calculating MLEs considering all parameters.

The E step: We will present the $(i+1)$st iteration. For each i, let $p^{(i)} = \{p_k^{(i)}\}_{k\geq 0}$ and $\theta^{(i)}$ be the offspring distribution and the control parameter, respectively, obtained in the ith iteration. Let $\mathcal{Z}_n^*|(\bar{\mathcal{Z}}_n, \{p^{(i)}, \theta^{(i)}\})$ be the distribution of the random vector $\mathcal{Z}_n^*$, (see [5.5]), given the sample $\bar{\mathcal{Z}}_n$ and the parameters values $p^{(i)}$ and $\theta^{(i)}$. For simplicity, we will use the notation $E_i^*[\cdot] = E_{\mathcal{Z}_n^*|(\bar{\mathcal{Z}}_n,\{p^{(i)},\theta^{(i)}\})}[\cdot]$. The log-likelihood function $\ell(p, \theta\,|\mathcal{Z}_n^*)$ given in [5.6] depends on the unobserved variables $Z_l(k)$ for $0 \leq l \leq n-1$ and $k \geq 0$. The expectation of the log-likelihood with respect to the distribution $\mathcal{Z}_n^*|(\bar{\mathcal{Z}}_n, \{p^{(i)}, \theta^{(i)}\})$ is

$$
\begin{aligned}
E_i^*[\ell(p,\theta \mid \mathcal{Z}_n^*, \overline{\mathcal{Z}}_n)] &= \Delta_{n-1}\log\theta - \sum_{l=0}^{n-1}\varphi(Z_l)\log b(\theta) \\
&\quad + \sum_{l=0}^{n-1}\sum_{k=0}^{\infty} E_i^*[Z_l(k)]\log p_k + E_i^*\,[K]\,.
\end{aligned}
$$

To calculate the above expectation, we need the distribution of $\mathcal{Z}_n^*$ given $\bar{\mathcal{Z}}_n$ with parameters $p^{(i)}$ and $\theta^{(i)}$. Let z_0, z_{l+1}, ϕ_l^*, $z_l(k) \in \mathbf{N}_0$, $0 \leq l \leq n-1$, $k \geq 0$, such that $z_{l+1} = \sum_{k=0}^{\infty} kz_l(k)$ and $\phi_l^* = \sum_{k=0}^{\infty} z_l(k)$. Define, for $l = 0, \ldots, n-1$,

$$
\begin{aligned}
A_l &:= \{Z_l(k) = z_l(k),\ k \geq 0\} \\
B_l &:= \{Z_{l+1} = z_{l+1}\} \\
C_l &:= \{\phi_l(Z_l) = \phi_l^*\}
\end{aligned}
$$

Since the individuals reproduce independently, and the control and offspring distributions are independent we have that

$$
\begin{aligned}
&P\Big(\bigcap_{l=0}^{n-1} A_l \,\Big|\, \bigcap_{l=0}^{n-1}(B_l\cap C_l)\cap\{Z_0 = z_0\}\Big) \\
&\quad = \prod_{l=0}^{n-1}\frac{P(A_l\cap B_l\cap C_l|\,Z_l = z_l)}{P(B_l\cap C_l\mid Z_l = z_l)}
\end{aligned}
$$

$$= \prod_{l=0}^{n-1} \frac{P(A_l \cap C_l | Z_l = z_l)}{P(B_l \cap C_l \mid Z_l = z_l)}$$

$$= \prod_{l=0}^{n-1} \frac{P(\phi_l(z_l) = \phi_l^*, \sum_{i=1}^{\phi_l(z_l)} \mathcal{I}\{X_{l,i} = k\} = z_l(k), \ k \geq 0)}{P(\sum_{i=1}^{\phi_l^*} X_{l,i} = z_{l+1}, \ \phi_l(z_l) = \phi_l^*)}$$

$$= \prod_{l=0}^{n-1} \frac{P(\sum_{i=1}^{\phi_l^*} \mathcal{I}\{X_{l,i} = k\} = z_l(k), \ k \geq 0)}{P(\sum_{i=1}^{\phi_l^*} X_{l,i} = z_{l+1})}$$

$$= \prod_{l=0}^{n-1} \frac{1}{P(\sum_{i=1}^{\phi_l^*} X_{l,i} = z_{l+1})} \cdot \frac{\phi_l^*!}{\prod_{k=0}^{\infty} z_l(k)!} \prod_{k=0}^{\infty} \left(p_k^{(i)}\right)^{z_l(k)} \quad [5.18]$$

Note that, even if the support of the offspring distribution is infinite, as long as z_{l+1} and ϕ_l^* are known, since $z_{l+1} = \sum_{k=0}^{\infty} k z_l(k)$ and $\phi_l^* = \sum_{k=0}^{\infty} z_l(k)$, only a finite number of coordinates of the sequence $\{z_l(k) : k \geq 0\}$ are non-zeros. From [5.18], it is clear that to obtain the distribution $\mathcal{Z}_n^* | (\bar{\mathcal{Z}}_n, \{p^{(i)}, \theta^{(i)}\})$, it is enough to know the distributions $(Z_l(k), k \geq 0) | (Z_l, \phi_l(Z_l), Z_{l+1}, \{p^{(i)}, \theta^{(i)}\})$ for each $l = 0, \ldots, n-1$. Let us fixed the generation index to be l and assume $Z_l = z_l$, $Z_{l+1} = z_{l+1}$ and $\phi_l(z_l) = \phi_l^*$. We need to determine the sample space of vector $(Z_l(k), k \geq 0)$ taking into account that its possible values $(z_l(k), \ k \geq 0)$ must satisfy the constrains $z_{l+1} = \sum_{k=0}^{\infty} k z_l(k)$ and $\phi_l^* = \sum_{k=0}^{\infty} z_l(k)$. After that, the corresponding probabilities must be obtained following the equation

$$\frac{1}{P(\sum_{i=1}^{\phi_l^*} X_{l,i} = z_{l+1})} \frac{\phi_l^*!}{\prod_{k=0}^{\infty} z_l(k)!} \prod_{k=0}^{\infty} \left(p_k^{(i)}\right)^{z_l(k)}.$$

To this end, it is enough to calculate them from a multinomial distribution for ϕ_l^* and $p^{(i)}$ and normalize the obtained probabilities. After that, it is straightforward to obtain the expected values $E_i^*[Z_l(k)]$ for $k \geq 0$. Note that this distribution does not depend on $\theta^{(i)}$ and hence it has no influence on the calculation of $E_i^*[Z_l(k)]$.

The M step: In the M step, we need to calculate the values of the parameters, that maximize the expectation of the complete log-likelihood, determined in the previous step. In other words, we have to find the values $p^{(i+1)} = \{p_k^{(i+1)}\}_{k \geq 0}$ and $\theta^{(i+1)}$ that maximize $E_i^*[\ell(p, \theta \mid \mathcal{Z}_n^*, \overline{\mathcal{Z}}_n)]$, subject to the constraints $p_k^{(i+1)} \geq 0$ for $k \geq 0$ and $\sum_{k=0}^{\infty} p_k^{(i+1)} = 1$.

Similarly to the calculation of the MLEs in the proof of theorem 5.1, we obtain for the values for p and θ in the $(i+1)$st iteration

$$
\begin{aligned}
p_k^{(i+1)} &= \frac{\sum_{l=0}^{n-1} E_i^*\left[Z_l(k)\right]}{\sum_{k=0}^{\infty}\sum_{l=0}^{n-1} E_i^*\left[Z_l(k)\right]} \\
&= \frac{\sum_{l=0}^{n-1} E_i^*\left[Z_l(k)\right]}{\sum_{l=0}^{n-1} E_i^*\left[\sum_{k=0}^{\infty} Z_l(k)\right]} \\
&= \frac{\sum_{l=0}^{n-1} E_i^*\left[Z_l(k)\right]}{\Delta_{n-1}}, \ k \geq 0
\end{aligned}
$$

and

$$
\theta^{(i+1)} = \mu^{-1}\left(\frac{\Delta_{n-1}}{\sum_{l=0}^{n-1}\varphi(Z_l)}\right).
$$

Intuitively, $p_k^{(i+1)}$ is the ratio of the average number of parents with k offspring (with respect to the probability distribution determined in the E step) and the total number of progenitors. Notice that $\theta^{(i+1)}$ does not depend on the iteration i because it is only based on the observed $\bar{\mathcal{Z}}_n$. Therefore, the algorithm reaches the value $\widehat{\theta}_n$, which never changes in the next iterations. Since $\theta^{(i)}$ plays no role in calculating $E_i^*[Z_l(k)]$, at iteration $(i+1)$ only $p_k^{(i)}$ is updated. Nevertheless, we include $\theta^{(i)}$ in the description of the algorithm, so it applies in both cases: when $\bar{\mathcal{Z}}_n$ is observed and when the sample is only $\{Z_0,\ldots,Z_n\}$. We discuss the latter case below.

The values $p^{(i+1)} = \{p_k^{(i+1)}\}_{k\geq 0}$ and $\theta^{(i+1)}$ obtained in the M step are next used to begin another E step and the procedure is repeated until a convergence criterion is satisfied. The values in the last iteration are denoted by $\widehat{p}_n^{(EM)} = \{\widehat{p}_{k,n}^{(EM)}\}_{k\geq 0}$ and $\widehat{\theta}_n^{(EM)}$, respectively. When $\bar{\mathcal{Z}}_n$ is observed, $\theta^{(i+1)}$ is not needed to begin another E step and obviously $\widehat{\theta}_n^{(EM)} = \widehat{\theta}_n$.

Sufficient conditions for the convergence of the algorithm are given in [MCL 08]. They involve differentiability of the expectation of the complete log-likelihood function. The sequence $\{(p^{(i)},\theta^{(i)})\}_{i\geq 0}$ converges to the MLE of (p,θ) based on the sample $\bar{\mathcal{Z}}_n$ provided that the likelihood function $\mathcal{L}(p,\theta|\,\bar{\mathcal{Z}}_n)$ is unimodal.

Once we obtain estimates for the parameter p, applying the EM algorithm, we can compute MLEs for m, σ^2 and $\mu(\theta)$ based on $\bar{\mathcal{Z}}_n$. Indeed,

$$\widehat{m}_n^{(EM)} = \sum_{k=0}^{\infty} k\widehat{p}_{k,n}^{(EM)}, \quad \widehat{\sigma}_n^{2(EM)} = \sum_{k=0}^{\infty} \left(k - \widehat{m}_n^{(EM)}\right)^2 \widehat{p}_{k,n}^{(EM)}, \text{ and}$$

$$\widehat{\mu}_n^{(EM)} = \mu(\widehat{\theta}_n^{(EM)}).$$

It is not difficult to see that $\widehat{m}_n^{(EM)} = \widehat{m}_n$ and $\widehat{\mu}_n^{(EM)} = \widehat{\mu}_n$. For example, for every $i = 0, 1, \ldots$,

$$\begin{aligned} m_n^{(i+1)} &= \sum_{k=0}^{\infty} kp_k^{(i+1)} \\ &= \frac{\sum_{k=0}^{\infty} k \sum_{l=0}^{n-1} E_i^*[Z_l(k)]}{\sum_{k=0}^{\infty} \sum_{l=0}^{n-1} E_i^*[Z_l(k)]} \\ &= \frac{\sum_{j=1}^{n} Z_j}{\Delta_{n-1}} \\ &= \widehat{m}_n. \end{aligned}$$

To summarize the procedure described so far, let us outline the steps of the EM algorithm for estimating p and θ, and consequently m, σ^2 and $\mu(\theta)$.

Step 0 Set $i = 0$ and choose $\theta^{(0)}$ and $0 \leq p_k^{(0)} \leq 1$, where $\sum_{k=0}^{\infty} p_k^{(0)} = 1$.

Step 1 *E step*. Based on $p^{(i)}$ and $\theta^{(i)}$

(a) determine $\mathcal{Z}_n^* | (\bar{\mathcal{Z}}_n, \{p^{(i)}, \theta^{(i)}\})$,

(b) calculate $E_i^*[\ell(p, \theta \; | \mathcal{Z}_n^*, \; \bar{\mathcal{Z}}_n)]$.

Step 2 *M step*. Calculate the values

$$(p^{(i+1)}, \theta^{(i+1)}) = \arg\max_{p,\theta} \; E_i^*[\ell(p, \theta \; | \mathcal{Z}_n^*, \bar{\mathcal{Z}}_n)].$$

`Step 3 If` $\max\{|p_k^{(i+1)} - p_k^{(i)}|,\ k = 0,1,\ldots,\ |\theta^{(i+1)} - \theta^{(i)}|\}$ `is less than a predetermine number (convergence criterion), then the algorithm halts and the final values are denoted by` $\widehat{p}_n^{(EM)}$ `and` $\widehat{\theta}_n^{(EM)}$`. Otherwise,` i `is incremented by one and steps 1-3 are repeated.`

Next, we will modify the EM algorithm outlined above to be able to estimate the parameters with assuming that **only the generation sizes can be observed**. Let us denote $\mathcal{Z}_n := \{Z_0, \ldots, Z_n\}$. We do not assume that the control functions are known exactly or the values $\phi_0(Z_0), \ldots, \phi_{n-1}(Z_{n-1})$ are observed. Some information on the kind of control, however, might be necessary, as it will be seen below. The expectation of the log–likelihood in the E step based on $\mathcal{Z}_n$ is

$$E_i[\ell(p,\theta|\mathcal{Z}_n^*,\ \mathcal{Z}_n)] = E_i[\Delta_{n-1}]\log\theta - \sum_{l=0}^{n-1}\varphi(Z_l)\log b(\theta)$$

$$+\sum_{l=0}^{n-1}\sum_{k=0}^{\infty} E_i[Z_l(k)]\log p_k + E_i[K],$$

where $E_i[\cdot]$ denotes the expectation with respect to $\mathcal{Z}_n^*|(\mathcal{Z}_n, \{p^{(i)}, \theta^{(i)}\})$. The calculation of this expectation requires knowledge of the distribution of $\mathcal{Z}_n^*$ given $\mathcal{Z}_n$ and the parameters $p^{(i)}$ and $\theta^{(i)}$. Similarly to [5.18], it can be proved that for $k = 0, 1, \ldots$

$$P(Z_l(k) = z_l(k),\ l = 0,\ldots,n-1, k \geq 0 \mid Z_0 = z_0, \ldots, Z_n = z_n) \qquad [5.19]$$

$$= \prod_{l=0}^{n-1} \frac{a_{z_l}(\phi_l^*)\theta^{\phi_l^*}}{P(Z_{l+1} = z_{l+1} \mid Z_l = z_l)} \frac{\phi_l^*!}{A(z_l,\theta)\prod_{k=0}^{\infty} z_l(k)!} \prod_{k=0}^{\infty}\left(p_k^{(i)}\right)^{z_l(k)},$$

where $\sum_{k=0}^{\infty} k z_l(k) = z_{l+1}$, and with $\phi_l^* = \sum_{k=0}^{\infty} z_l(k)$, $0 \leq l \leq n-1$. In order to use [5.19] to determine the distribution $\mathcal{Z}_n^*|(\mathcal{Z}_n, \{p^{(i)}, \theta^{(i)}\})$, we need to know the distribution of $(Z_l(k), k \geq 0) \mid (Z_l, Z_{l+1}, \{p^{(i)}, \theta^{(i)}\})$, $0 \leq l \leq n-1$. First, fixing l, we consider the possible values for ϕ_l^* obtained from the control distribution with parameters $\theta^{(i)}$ and z_l (here, we need information about the kind of control distribution). Then, for each fixed value ϕ_l^*, we calculate the sample space of the vector $(Z_l(k), k \geq 0)$ taking into account that its possible values $(z_l(k), k \geq 0)$ must verify the constrains

$z_{l+1} = \sum_{k=0}^{\infty} k z_l(k)$ and $\phi_l^* = \sum_{k=0}^{\infty} z_l(k)$. Finally, their corresponding probabilities equal the product of probabilities from a multinomial distribution with parameters ϕ_l^* and $p^{(i)}$ and the probability that the control distribution takes on the value ϕ_l^* (suitably normalized).

The values of p and θ, which maximize the expectation [5.19], subject to the constraints $p_k^{(i+1)} \geq 0$, $k \geq 0$, and $\sum_{k=0}^{\infty} p_k^{(i+1)} = 1$, are

$$\begin{aligned} p_k^{(i+1)} &= \frac{\sum_{l=0}^{n-1} E_i\left[Z_l(k)\right]}{\sum_{k=0}^{\infty}\sum_{l=0}^{n-1} E_i\left[Z_l(k)\right]} \\ &= \frac{\sum_{l=0}^{n-1} E_i\left[Z_l(k)\right]}{\sum_{l=0}^{n-1} E_i\left[\sum_{k=0}^{\infty} Z_l(k)\right]} \\ &= \frac{\sum_{l=0}^{n-1} E_i\left[Z_l(k)\right]}{E_i\left[\Delta_{n-1}\right]}, \quad k \geq 0 \end{aligned}$$

and

$$\theta^{(i+1)} = \mu^{-1}\left(\frac{E_i\left[\Delta_{n-1}\right]}{\sum_{l=0}^{n-1} \varphi(Z_l)}\right).$$

The values $p^{(i+1)} = \{p_k^{(i+1)}\}_{k\geq 0}$ and $\theta^{(i+1)}$ obtained in the M step are next used to begin another E step and the procedure is repeated until the convergence criterion is satisfied. Once the estimates $\widetilde{p}_n^{(EM)} = \{\widetilde{p}_{k,n}^{(EM)}\}_{k\geq 0}$ and $\widetilde{\theta}_n^{(EM)}$ are calculated, we can find estimates for m, σ^2 and $\mu(\theta)$ based on $\mathcal{Z}_n$ defined as follows:

$$\widetilde{m}_n^{(EM)} = \sum_{k=0}^{\infty} k\widetilde{p}_{k,n}^{(EM)}, \quad \widetilde{\sigma}_n^{2(EM)} = \sum_{k=0}^{\infty}\left(k - \widetilde{m}_n^{(EM)}\right)^2 \widetilde{p}_{k,n}^{(EM)},$$

$$\widetilde{\mu}_n^{(EM)} = \mu(\widetilde{\theta}_n^{(EM)}).$$

5.1.3. *Simulated example*

We will illustrate the behavior of the proposed MLEs with a simulated example. Consider a CB process with offspring distribution

$$p_0 = 0.1084, \qquad p_1 = 0.2709, \qquad p_2 = 0.3386, \qquad p_3 = 0.2822$$

and binomial control variables $\phi_n(k)$ with parameters k and 0.6. Note that in this setting $\varphi(\cdot)$ is the identity function, so that no prior deterministic control is assumed. Under these assumptions, we have

$$m = 1.7946, \quad \sigma^2 = 0.9443, \quad \theta = 1.5, \quad \mu(\theta) = 0.6, \quad \tau_m = 1.0767.$$

CB process of this kind can be used as a model of the evolution of a population in the presence of predators.

If the population size is given, then the knowledge of either θ or $\mu(\theta)$ is sufficient to determine the control distribution. Therefore, without loss of generality, we will focus on estimating the migration parameter $\mu(\theta)$. The subject matter process is a supercritical CB process with supercritical offspring mean, i.e. $m > 1$. Note that 40% of the individuals are eliminated (e.g. victims of predators) and do not participate in the reproduction process for the next generation.

We simulate the first 30 generations of this process starting with a single ancestor $Z_0 = 1$. Denote the three different samples schemes: entire family tree, individuals and progenitors, and population size only by z_{30}^*, $\overline{z}_{30}$ and z_{30}, respectively. Figure 5.1 shows both number of individuals and progenitors in the first 30 generations of a trajectory of the process (see data in Appendix 4). Clearly the reproduction dominates over the control. As a result, the population size increases despite the emigration/predators. Thus, this simulated realization seems to belong to the set $\{Z_n \to \infty\}$. Note that, the results in Chapter 3 and 4 imply that, on the set $\{Z_n \to \infty\}$, the process grows exponentially with rate τ_m and thus the assumptions in [5.7] hold.

First, we calculate the MLEs and their approximate 95% confidence intervals based on the entire family tree, z_{30}^*. The estimates are given in Table 5.1. Figures 5.2–5.4 show their behavior over the course of the first 30 generations. All estimates approach the true values of the parameters as it is expected by theorems 5.2 and 5.3.

Next we apply the EM algorithm in the two sample schemes: observing the total number of individuals and progenitors in each generation and observing the generation sizes only. In both cases, we assume no other information is available about the offspring distribution but the maximum number of offspring per progenitor. We start the algorithm with a Uniform distribution on $\{0, 1, 2, 3\}$ and $\theta^{(0)} = 1/2$. The maximum number of offspring per progenitor in an animal population is a datum that is likely to be known once the reproductive cycle of the corresponding species is understood. Even if this information is unavailable, one can use various

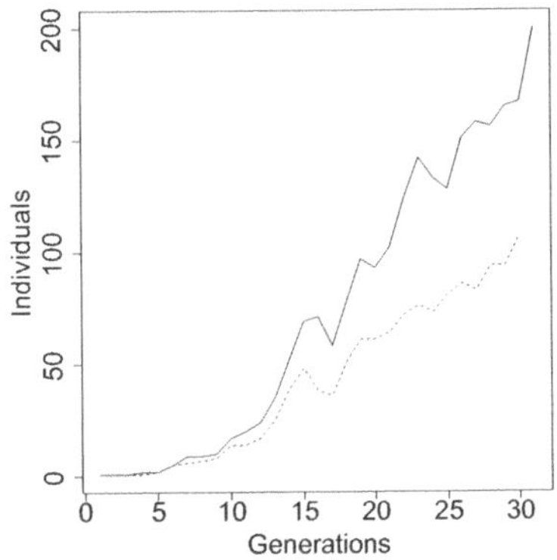

Figure 5.1. *Evolution of the number of individuals (solid line) and the number of progenitors (dashed line). Reprinted from [GON 16b]. Copyright (2016), with permission from Elsevier*

	Parameters							
	p_0	p_1	p_2	p_3	m	σ^2	$\mu(\theta)$	τ_m
MLE based on z_{30}^*	0.1027	0.2765	0.3389	0.2820	1.8002	0.9293	0.6087	1.0957
EM based on $\overline{z}_{30}$	0.1211	0.2528	0.3308	0.2953	1.8002	0.9927	0.6087	1.0957
EM based on z_{30}	0.1299	0.3083	0.3283	0.2335	1.6653	0.9496	0.6579	1.0957
TRUE VALUES	0.1084	0.2709	0.3386	0.2822	1.7946	0.9443	0.6000	1.0767

Table 5.1. *Estimates based on the samples z_{30}^*, $\overline{z}_{30}$ and z_{30}. Reprinted from [GON 16b]. Copyright (2016), with permission from Elsevier*

reasonable values for this maximum and use Akaike information criterion in choosing the optimal value (see [GON 16b]).

Using the data on individuals plus progenitors, we ran the algorithm until attaining a difference between two consecutive iterations smaller than 10^{-6}. This convergence occurred from iteration 733 onwards. The resulting estimates are given in Table 5.1. After achieving the desired convergence, we ran the procedure using samples $\overline{z}_j$, $j = 1, \ldots, 30$, to assess the consistency of the estimates. Figures 5.5 and 5.6 show the evolution of these estimates obtained after convergence of the EM algorithm, and based on the samples $\overline{z}_j$, $j =$

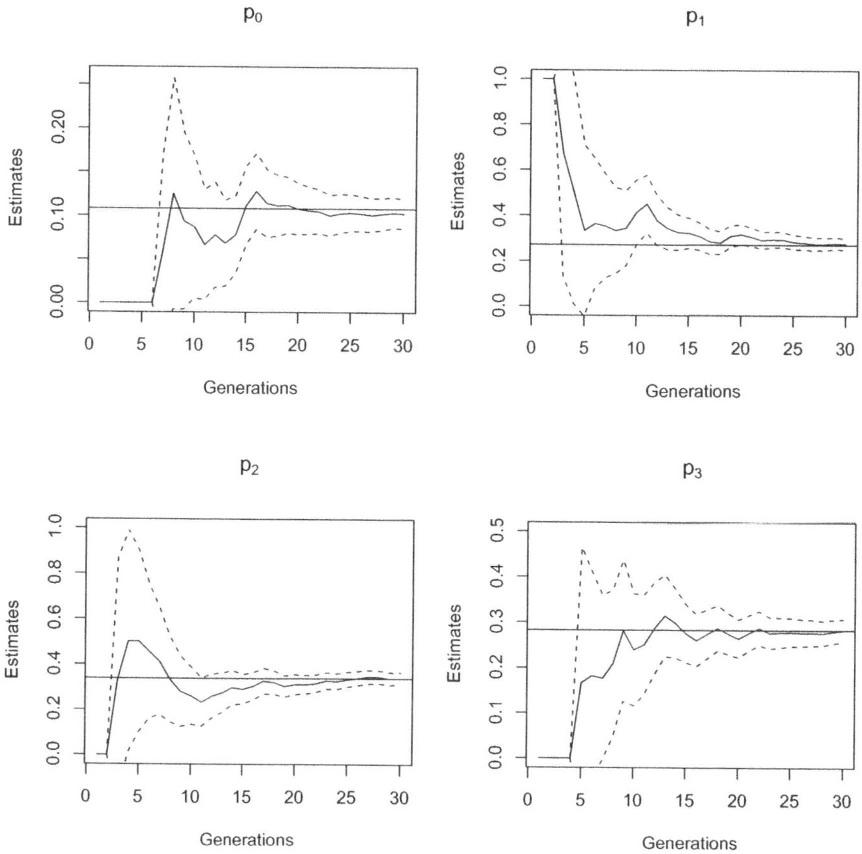

Figure 5.2. *Evolution of the estimates given by $\widehat{p}_0$, $\widehat{p}_1$, $\widehat{p}_2$ and $\widehat{p}_3$ (solid line), and their approximate 95% confidence intervals (dashed line). The horizontal line represents the true value of each parameter. Reprinted from [GON 16b]. Copyright (2016), with permission from Elsevier*

$1, \ldots, 30$ (dashed lines), together with MLEs based on the entire family tree, for the parameters p_k, $k = 0, 1, 2, 3$, m, σ^2, and $\mu(\theta)$. As was pointed out in the previous section, the EM algorithm is not needed for the MLEs of m, θ and $\mu(\theta)$, so the estimates based on the total number of individuals and progenitors in each generation are repeated in the table.

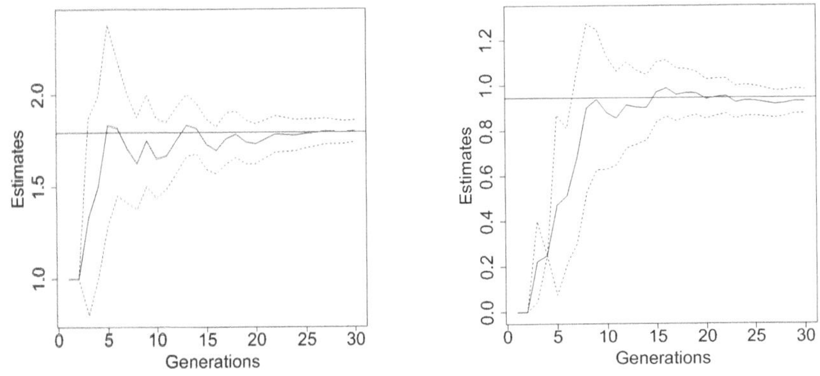

Figure 5.3. *Evolution of the estimates given by $\widehat{m}_n$ (left) and $\widehat{\sigma}^2_n$ (right) over the course of the generations (solid line) and their approximate 95% confidence intervals (dashed line). The horizontal line represents the true value of each parameter. Reprinted from [GON 16b]. Copyright (2016), with permission from Elsevier*

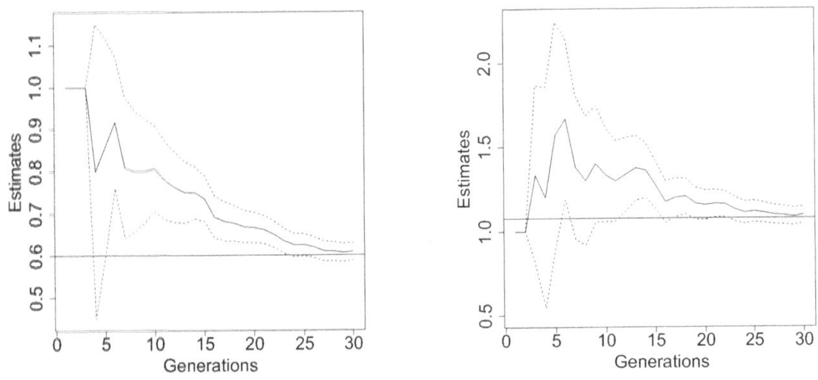

Figure 5.4. *Evolution of the estimates given by $\widehat{\mu}_n$ (left) and $\widehat{\tau}_m$ (right) over the course of the generations (solid line) and their approximate 95% confidence intervals (dashed line). The horizontal line represents the true value of each parameter. Reprinted from [GON 16b]. Copyright (2016), with permission from Elsevier*

We also applied the EM algorithm using the observations on the population sizes, $\overline{z}_{30}$, only. Recall that it is necessary in this case to know the

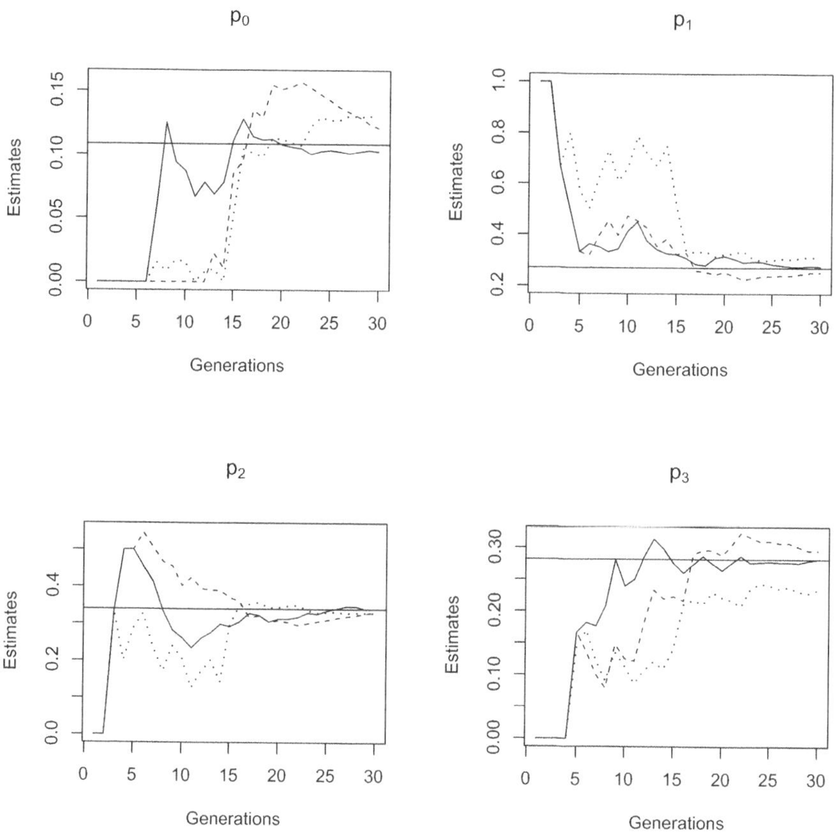

Figure 5.5. *Evolution of the estimates given by $\widehat{p}$ (solid lino), $\widehat{p}^{(EM)}$ (dashed line) and $\widetilde{p}^{(EM)}$ (dotted line). Reprinted from [GON 16b]. Copyright (2016), with permission from Elsevier*

kind of control distribution. This information can be based on general considerations about population's development. For instance, in the presence of predators a binomial control distribution would be justified. In the simulation, we observed convergence from iteration 1164 onwards (with precision of 10^{-6}). The estimates of the parameters are listed in Table 5.1 and their evolution is plotted in Figures 5.5 and 5.6. It is remarkable that all estimates approach the respective true parameter as the generation number increases.

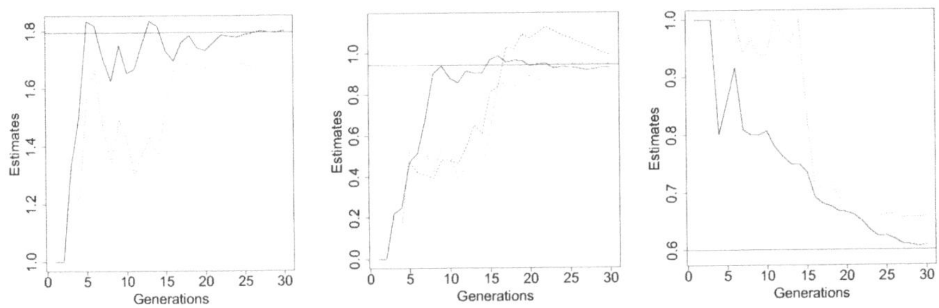

Figure 5.6. *Evolution of the estimates for m (left), σ^2 (middle) and $\mu(\theta)$ (right) based on the entire family tree (solid line), on the total number of individuals and progenitors (dashed line) (for estimates of m and $\mu(\theta)$ dashed and solid lines coincide) and on the total number of individuals (dotted line). The horizontal line represents the true value of each parameter. Reprinted from [GON 16b]. Copyright (2016), with permission from Elsevier*

For further investigations on the influence of the initial values of $(p^{(0)}, \theta^{(0)})$ on the convergence of the algorithms, we refer to [GON 16b]. The same reference provides a bootstrap approximation for the sampling distributions of the estimators. Finally, in [GON 16b] it is also discussed the order of the computational complexity of each iteration in the EM algorithms.

5.2. Conditional weighted least squares estimation

In this section, using the conditional weighted least squares (CWLS) method, we will construct estimators for the offspring mean. We will also study the asymptotic behavior of the estimators in all three modes of criticality. Note that conditions about the distributions of the control variables are not needed.

Recall from proposition 3.5 in Chapter 3 that for $\mathcal{F}_n = \sigma(Z_0, Z_1, \cdots, Z_n)$

$$E[Z_n|\mathcal{F}_{n-1}] = m\varepsilon(Z_{n-1}), \tag{5.20}$$

$$Var[Z_n|\mathcal{F}_{n-1}] = \sigma^2\varepsilon(Z_{n-1}) + m^2\nu^2(Z_{n-1}), \tag{5.21}$$

where $\varepsilon(k) = E[\phi_n(k)]$ and $\nu^2(k) = Var[\phi_n(k)]$. Equation [5.20] yields

$$Z_n = m\varepsilon(Z_{n-1}) + \tilde{\alpha}_n, \qquad n = 1, 2, \ldots, \tag{5.22}$$

where the error term $\tilde{\alpha}_n$ has $E[\tilde{\alpha}_n|\mathcal{F}_{n-1}] = 0$. It can be seen that since $Var[\tilde{\alpha}_n|\mathcal{F}_{n-1}] = \sigma^2\varepsilon(Z_{n-1}) + m^2\nu^2(Z_{n-1})$, the conditional least squares estimators based on [5.22] are not efficient. To overcome this, let us rewrite the model [5.22] as

$$\frac{Z_n}{\sqrt{\varepsilon(Z_{n-1})+1}} = \frac{m\varepsilon(Z_{n-1})}{\sqrt{\varepsilon(Z_{n-1})+1}} + \alpha_n, \qquad n = 1, 2, \ldots, \qquad [5.23]$$

where $\alpha_n = \tilde{\alpha}_n / \sqrt{\varepsilon(Z_{n-1})+1}$ and by [5.20] and [5.21] we have

$$E[\alpha_n|\mathcal{F}_{n-1}] = 0, \quad \text{and} \quad Var[\alpha_n|\mathcal{F}_{n-1}] = \frac{\sigma^2\varepsilon(Z_{n-1}) + m^2\nu^2(Z_{n-1})}{\varepsilon(Z_{n-1})+1}.$$

As we will show below, under certain regularity conditions on $\{\varepsilon(k)\}_{k\geq 0}$ and $\{\nu^2(k)\}_{k\geq 0}$, the conditional variance of the error term in [5.23] is bounded, which makes the CWLS estimators based on [5.23] better than those based on [5.22]. It is not difficult to see that [5.23] yields the CWLS estimators for the offspring mean given by

$$\hat{m}_n := \frac{1}{A_n}\sum_{i=1}^{n} \frac{Z_i\varepsilon(Z_{i-1})}{\varepsilon(Z_{i-1})+1}, \qquad \text{where} \quad A_n := \sum_{i=1}^{n} \frac{\varepsilon^2(Z_{i-1})}{\varepsilon(Z_{i-1})+1}. \qquad [5.24]$$

Next, we study the **limiting distributions of** $\hat{m}_n$ **as** $n \to \infty$ for subcritical, critical and supercritical CB processes separately.

5.2.1. *Subcritical processes*

Under hypotheses of theorem 4.1 in Chapter 4, it was proved that a subcritical CB process, $\{Z_n\}_{n\geq 0}$, converges in distribution to a positive, finite and non-degenerate random variable Z. We will also assume that for some $\delta > 0$

$$E[|\phi_0(Z) - \varepsilon(Z)|^{2+\delta}] < \infty. \qquad [5.25]$$

THEOREM 5.4.– *Let* $\{Z_n\}_{n\geq 0}$ *be a subcritical CB process satisfying the assumptions of theorem 4.1 and* [5.25]. *Then*

$$\sqrt{A_n}\,(\hat{m}_n - m) \xrightarrow{d} N(0, v^2) \quad \text{as} \quad n \to \infty,$$

where

$$v^2 := \frac{1}{E\left[\dfrac{\varepsilon^2(Z)}{\varepsilon(Z)+1}\right]}\left(m^2 E\left[\frac{\varepsilon^2(Z)\nu^2(Z)}{(\varepsilon(Z)+1)^2}\right] + \sigma^2 E\left[\frac{\varepsilon^3(Z)}{(\varepsilon(Z)+1)^2}\right]\right)$$

Proof. Write

$$\sqrt{A_n}\,(\hat{m}_n - m) = \sqrt{A_n}\left(\frac{1}{A_n}\sum_{i=1}^{n}\frac{Z_i\varepsilon(Z_{i-1})}{\varepsilon(Z_{i-1})+1} - m\right)$$

$$= \frac{\dfrac{1}{\sqrt{n}}\displaystyle\sum_{i=1}^{n}\frac{(Z_i - m\varepsilon(Z_{i-1}))\varepsilon(Z_{i-1})}{\varepsilon(Z_{i-1})+1}}{\sqrt{\dfrac{A_n}{n}}}.$$

Denote

$$Y_{ni} := \frac{(Z_i - m\varepsilon(Z_{i-1}))\varepsilon(Z_{i-1})}{\sqrt{n}(\varepsilon(Z_{i-1})+1)} \quad \text{and} \quad v_n^2 := \sum_{i=1}^{n} E\left[Y_{ni}^2|\mathcal{G}_{n,i-1}\right],$$

where $\mathcal{G}_{n,i-1} \equiv \mathcal{F}_{i-1}$ as defined previously. By **Ergodic theorem**, we have

$$\lim_{n\to\infty}\frac{1}{n}A_n = E\left[\frac{\varepsilon^2(Z)}{\varepsilon(Z)+1}\right] \qquad \text{a.s.}$$

and

$$\lim_{n\to\infty} v_n^2 = m^2 E\left[\frac{\varepsilon^2(Z)\nu^2(Z)}{(\varepsilon(Z)+1)^2}\right] + \sigma^2 E\left[\frac{\varepsilon^3(Z)}{(\varepsilon(Z)+1)^2}\right] \qquad \text{a.s.}$$

Since $\lim_{n\to\infty} v_n^2$ is a constant, the claim of the theorem will follow from theorem A1.4 in Appendix 1, once we show that for all $\epsilon > 0$,

$$\lim_{n\to\infty}\sum_{i=1}^{n} E\left[Y_{ni}^2\mathcal{I}\{|Y_{ni}| > \epsilon\} \mid \mathcal{G}_{n,i-1}\right] = 0 \qquad \text{in probability,}$$

where recall $\mathcal{I}A$ is the indicator of set A. For this to be true, it is sufficient to show that for $\delta \in (0, 2)$

$$\lim_{n\to\infty}\sum_{i=1}^{n} E\left[|Y_{ni}|^{2+\delta}\,|\mathcal{G}_{n,i-1}\right] = 0 \qquad \text{in probability.} \qquad [5.26]$$

Indeed, there exists $K_\delta > 0$, such that

$$E\left[|Y_{ni}|^{2+\delta} \mid \mathcal{G}_{n,i-1}\right]$$

$$\leq \frac{2^{\delta+1}}{n^{1+\delta/2}}\left(K_\delta E\left[\phi_0(Z_{i-1})^{1+\delta/2}\right] + m^{2+\delta} E\left[|\phi_0(Z_{i-1}) - \varepsilon(Z_{i-1})|^{2+\delta}\right]\right).$$

Now, [5.26] follows taking into account assumption [5.25] and the Ergodic theorem. ∎

5.2.2. *Critical processes*

In this section, we will find the limit distribution of the CWLS estimator in the critical case, when $P(Z_n \to \infty) > 0$ and the following conditions hold

$$\tau_m(k) = 1 + \frac{\gamma}{k} + o\left(\frac{1}{k}\right) \qquad \text{as } k \to \infty, \quad -\infty < \gamma < \infty, \qquad [5.27]$$

$$\lim_{k\to\infty} \frac{\nu^2(k)}{k} = 0. \qquad [5.28]$$

Sufficient conditions for $P(Z_n \to \infty) > 0$ for critical CB processes were obtained in Chapter 3. For instance, if [5.27] and [5.28] hold with $\gamma > \sigma^2/(2m)$, then $P(Z_n \to \infty) > 0$ under an additional assumption on conditional moments (see theorem 3.4).

Recall the notation introduced in section 4.2.3 in Chapter 4. Let $W_n(t) = Z_{[nt]}/n$ for $t \geq 0$. As a particular case of theorem 4.4 the following result holds (for a proof, see [SRI 07]).

THEOREM 5.5.– *Let* $\{Z_n\}_{n\geq 0}$ *be a critical CB process, such that* [5.27] *and* [5.28] *hold. Then,* $W_n \xrightarrow{\mathcal{D}} W$ *as* $n \to \infty$*, where* W *is the diffusion process with generator*

$$Af(x) = \gamma f'(x) + \frac{x\sigma^2}{2m} f''(x), \qquad f \in C_c^\infty[0,\infty).$$

The process W *is the (unique) solution of the stochastic differential equation*

$$dW(t) = \gamma dt + \sqrt{\frac{\sigma^2}{m} W(t)}\, dB(t), \qquad t \in [0,\infty),$$

where B *is a standard Wiener process.*

The following proposition, whose proof can be read in [SRI 07] stated as lemma 2, is obtained as a consequence of a general version of the continuous mapping theorem (see Appendix 2).

PROPOSITION 5.2.– *Let* $\{Z_n\}_{n\geq 1}$ *be a critical CB process, such that* [5.27] *and* [5.28] *hold. Then, as* $n \to \infty$

$$\frac{W_n(1) - W_n(0) - \gamma}{\sqrt{\dfrac{1}{m}\displaystyle\int_0^1 \tau_m(Z_{[nt]})W_n(t)\,dt}} \xrightarrow{d} \frac{W(1) - W(0) - \gamma}{\sqrt{\dfrac{1}{m}\displaystyle\int_0^1 W(t)\,dt}}.$$

Now, we are in position to obtain the limit of the CWLS estimator in the critical case.

THEOREM 5.6.– *Let* $\{Z_n\}_{n\geq 1}$ *be a critical CB process, such that* [5.27] *and* [5.28] *hold. Then, as* $n \to \infty$

$$\sqrt{A_n}\,(\hat{m}_n - m) \xrightarrow{d} \frac{W(1) - W(0) - \gamma}{\sqrt{\dfrac{1}{m}\displaystyle\int_0^1 W(t)\,dt}}.$$

Proof. For simplicity assume $P(Z_n \to \infty) = 1$. Denote

$$U_n = \frac{1}{\sum_{j=1}^{n} \varepsilon(Z_{j-1})}.$$

With this notation, we can write

$$\sqrt{A_n}(\hat{m}_n - m) = \frac{1}{\sqrt{A_n}}\left(\sum_{i=1}^{n}(Z_i - m\varepsilon(Z_{i-1})) - \sum_{i=1}^{n}\frac{Z_i - m\varepsilon(Z_{i-1})}{1 + \varepsilon(Z_{i-1})}\right)$$

$$= \frac{B_n - C_n}{\sqrt{1 - D_n}},$$

where

$$B_n := \sqrt{U_n}\sum_{i=1}^{n}(Z_i - m\varepsilon(Z_{i-1})),\ C_n := \sqrt{U_n}\sum_{i=1}^{n}\frac{Z_i - m\varepsilon(Z_{i-1})}{1 + \varepsilon(Z_{i-1})},$$

$$D_n := 1 - A_n U_n.$$

Now we have

$$B_n = \frac{\dfrac{1}{n}\displaystyle\sum_{i=1}^{n}(Z_i - Z_{i-1}) + \dfrac{1}{n}\sum_{i=1}^{n}(Z_{i-1} - m\varepsilon(Z_{i-1}))}{\dfrac{1}{n}\sqrt{\dfrac{1}{U_n}}}$$

$$= \frac{\dfrac{Z_n - Z_0}{n} + \dfrac{1}{n}\displaystyle\sum_{i=1}^{n}(Z_{i-1} - m\varepsilon(Z_{i-1}))}{\dfrac{1}{n}\sqrt{\dfrac{1}{U_n}}}$$

$$= \frac{W_n(1) - W_n(0) - \displaystyle\sum_{i=1}^{n}\int_{(i-1)/n}^{i/n} Z_{[nt]}\left(\frac{m\varepsilon(Z_{[nt]})}{Z_{[nt]}} - 1\right) dt}{\sqrt{\dfrac{1}{n}\displaystyle\sum_{i=1}^{n}\int_{(i-1)/n}^{i/n} \varepsilon(Z_{[nt]})\, dt}}$$

and finally

$$B_n = \frac{W_n(1) - W_n(0) - \displaystyle\int_0^1 Z_{[nt]}\left(\tau_m(Z_{[nt]}) - 1 - \frac{\gamma}{Z_{[nt]}}\right) dt - \gamma}{\sqrt{\dfrac{1}{m}\displaystyle\int_0^1 \frac{m\varepsilon(Z_{[nt]})}{Z_{[nt]}} W_n(t)\, dt}}.$$

To complete the proof, we need to show the following.

a) $\lim_{n\to\infty} D_n = 0$ in probability.

b) $\lim_{n\to\infty} C_n = 0$ a.s.

c)

$$\lim_{n\to\infty} \int_0^1 Z_{[nt]}\left(\tau_m(Z_{[nt]}) - 1 - \frac{\gamma}{Z_{[nt]}}\right) dt = 0 \quad \text{a.s.}$$

d)

$$\lim_{n\to\infty} B_n = \frac{W(1) - W(0) - \gamma}{\sqrt{\dfrac{1}{m}\displaystyle\int_0^1 W(t)\, dt}} \quad \text{in distribution.}$$

To verify a), observe that

$$|D_n| \leq nU_n = \frac{1}{n} \frac{1}{\frac{1}{n^2}\sum_{i=1}^{n} \varepsilon(Z_{i-1})} \qquad [5.29]$$

and, using the arguments in the proof of proposition 5.2,

$$\lim_{n\to\infty} \frac{1}{n^2} \sum_{i=1}^{n} \varepsilon(Z_{i-1}) = \frac{1}{m} \int_0^1 W(t)\, dt \qquad \text{in distribution} \qquad [5.30]$$

Next, consider b). Setting

$$V_n := \sum_{i=1}^{n} \frac{1}{1+\varepsilon(Z_{i-1})},$$

we rewrite C_n as

$$C_n = V_n \sqrt{U_n}\, \frac{1}{V_n} \sum_{i=1}^{n} \frac{Z_i - m\varepsilon(Z_{i-1})}{1+\varepsilon(Z_{i-1})}.$$

It can be proved (see lemma 3 in [SRI 07]) that $V_n \to \infty$ a.s. Now clearly,

$$V_n \sqrt{U_n} \leq n\sqrt{U_n} \xrightarrow{d} \frac{1}{\sqrt{\frac{1}{m} \int_0^1 W(t)\, dt}}, \qquad [5.31]$$

where the limit follows from [5.30]. It remains to consider

$$\frac{1}{V_n} \sum_{i=1}^{n} \frac{Z_i - m\varepsilon(Z_{i-1})}{1+\varepsilon(Z_{i-1})}.$$

It can be verified that

$$\sum_{i=1}^{n} \frac{Z_i - m\varepsilon(Z_{i-1})}{1+\varepsilon(Z_{i-1})}$$

is a martingale. Therefore, considering $\mathcal{F}_i = \sigma(Z_0, \ldots, Z_i),\ i \geq 0$, as before, by theorem A1.3 in Appendix 1,

$$\sum_{i=1}^{\infty} \frac{E\left[\left(\frac{Z_i - m\varepsilon(Z_{i-1})}{1+\varepsilon(Z_{i-1})}\right)^2 \Big| \mathcal{F}_{i-1}\right]}{V_i^2} < \infty$$

implies

$$\lim_{n\to\infty} \frac{1}{V_n} \sum_{i=1}^{n} \frac{Z_i - m\varepsilon(Z_{i-1})}{1+\varepsilon(Z_{i-1})} = 0 \quad \text{a.s.} \qquad [5.32]$$

Using [5.27] and [5.28], we obtain

$$E\left[\left(\frac{Z_i - m\varepsilon(Z_{i-1})}{1+\varepsilon(Z_{i-1})}\right)^2 \Big| \mathcal{F}_{i-1}\right] = \frac{m^2\nu^2(Z_{i-1}) + \sigma^2\varepsilon(Z_{i-1})}{(1+\varepsilon(Z_{i-1}))^2}$$
$$\leq \frac{C_0}{1+\varepsilon(Z_{i-1})},$$

where C_0 is a generic constant. Therefore,

$$\sum_{i=1}^{\infty} \frac{E\left[\left(\frac{Z_i - m\varepsilon(Z_{i-1})}{1+\varepsilon(Z_{i-1})}\right)^2 \Big| \mathcal{F}_{i-1}\right]}{V_i^2} \leq C_0 \sum_{i=1}^{\infty} \frac{(1+\varepsilon(Z_{i-1}))^{-1}}{V_i^2}$$
$$= C_0 \sum_{i-1}^{\infty} \frac{(V_i - V_{i-1})}{V_i^2}$$
$$< \infty.$$

Combining [5.31] and [5.32], we obtain $\lim_{n\to\infty} C_n = 0$ a.s.

As for the proof of c), for any fixed $\omega \in \{Z_n \to \infty\}$, define a sequence of random variables $\{g_n\}_{n\geq 1}$, such that $g_n : ([0,1], \mathcal{B}[0,1], \lambda) \longrightarrow (\mathbf{R}, \mathcal{B}(\mathbf{R}))$, where λ is the Lebesgue measure, and $\mathcal{B}[0,1]$ and $\mathcal{B}(\mathbf{R})$ the Borel σ-algebra of the corresponding sets, such that

$$g_n(t) = Z_{[nt]}(\omega)\left(\tau_m(Z_{[nt]}(\omega)) - 1 - \frac{\gamma}{Z_{[nt]}(\omega)}\right).$$

Then, it follows by [5.27] that $\lim_{n\to\infty} g_n = 0$ a.s. with respect to λ. Therefore, by the dominated convergence theorem,

$$\lim_{n\to\infty} \int_0^1 Z_{[nt]}(\omega) \left(\tau_m(Z_{[nt]}(\omega)) - 1 - \frac{\gamma}{Z_{[nt]}(\omega)} \right) dt = 0.$$

Using c) and proposition 5.2, we obtain d) and hence the result is proved. ∎

5.2.3. *Supercritical processes*

Consider now supercritical CB processes with condition [5.7], i.e. such that $\lim_{k\to\infty} \tau_m(k) = \tau_m > 1$,

$$P(Z_n \to \infty) > 0 \quad \text{and} \quad \lim_{n\to\infty} W_n = W \quad \text{a.s.},$$

where $W_n = Z_n/\tau_m^n$ and $P(W > 0) > 0$. It follows from [5.7] (iii) and proposition 5.1, 1) that on $\{Z_n \to \infty\}$,

$$\lim_{n\to\infty} \frac{\varepsilon(Z_n)}{\tau_m^n} = \tau W \qquad \text{a.s.,}$$

Therefore, using Cesàro's lemma, on $\{Z_n \to \infty\}$,

$$\lim_{n\to\infty} \frac{1}{\tau_m^n} A_n = \frac{\tau W}{\tau_m - 1} \quad \text{a.s.} \quad \text{where as before} \quad A_n = \sum_{i=1}^{n} \frac{\varepsilon^2(Z_{i-1})}{\varepsilon(Z_{i-1}) + 1}. \qquad [5.33]$$

Below we will find the limit of the CWLS estimator for the mean of a supercritical CB process, assuming [5.7] and

$$\lim_{k\to\infty} \frac{\nu^2(k)}{k} = 0. \qquad [5.34]$$

THEOREM 5.7.– *Let* $\{Z_n\}_{n\geq 0}$ *be a supercritical CB process. If* [5.7] *and* [5.34] *hold, then*

$$\sqrt{A_n}\,(\hat{m}_n - m) \xrightarrow{d} N\left(0, \sigma^2\right) \quad \text{as} \quad n \to \infty.$$

Proof. For simplicity will assume $P(Z_n \to \infty) = 1$. The scheme of the proof is similar to that in theorem 5.3. We rewrite

$$\sqrt{A_n}\,(\hat{m}_n - m) = \frac{1}{\sqrt{A_n}} \sum_{i=1}^{n} (Z_i - m\varepsilon(Z_{i-1})) \left(\frac{\varepsilon(Z_{i-1})}{\varepsilon(Z_{i-1}) + 1}\right) = \frac{1}{\sqrt{A_n}} ((I) + (II)), \quad [5.35]$$

where

$$(I) := \sum_{i=1}^{n} \left(\eta_{ni} \left(\sqrt{Z_{i-1} + 1} - \sqrt{W \tau_m^{i-1}}\right)\right),$$

$$(II) := \sqrt{W} \sum_{i=1}^{n} \eta_{ni} \sqrt{\tau_m^{i-1}},$$

and

$$\eta_{ni} := \frac{\varepsilon(Z_{i-1})}{\varepsilon(Z_{i-1}) + 1} \frac{Z_i - m\varepsilon(Z_{i-1})}{\sqrt{Z_{i-1} + 1}}, \qquad i = 1, 2, \ldots, n.$$

Consider (II) above. It can be shown that $(\eta_{n1}, \eta_{n2}, \ldots, \eta_{nn})$ converges in distribution to a multivariate normal random vector with mean vector null and covariance matrix $\tau\sigma^2 I_n$, where I_n is the identity matrix of size n. Indeed, to prove this, following the proof of theorem 1 in [HEY 71] (or [SHE 03]), it is enough to check that, for all sequences $z_n \to \infty$, as $n \to \infty$,

$$\frac{1}{\sqrt{z_n + 1}} \left(\sum_{j=1}^{\phi_n(z_n)} (X_{nj} - m) + (m\phi_n(z_n) - m\varepsilon(z_n))\right) \xrightarrow{d} N(0, \tau\sigma^2) \quad [5.36]$$

Using the central limit theorem for randomly stopped sums (Doeblin–Anscombe's theorem, see theorem A2.2 in Appendix 2), we obtain that the first summand in [5.36] converges in distribution to $N(0, \tau\sigma^2)$, as $n \to \infty$. Also, it is not difficult to check, using Markov's inequality and [5.34], that the second summand in [5.36] converges in probability to 0, as $n \to \infty$. Hence [5.36] holds.

The multivariate normal limit result yields

$$\sum_{i=1}^{n} \frac{\eta_{ni}}{\sqrt{\tau_m^i}} \xrightarrow{d} N\left(0, \tau\sigma^2(\tau_m - 1)^{-1}\right) \quad \text{as} \quad n \to \infty.$$

Therefore,

$$\frac{1}{\sqrt{A_n}}(II) \xrightarrow{d} N\left(0, \sigma^2\right) \quad \text{as} \quad n \to \infty. \qquad [5.37]$$

Now, to investigate the limiting behavior of (I), similarly to the proof of theorem 5.3 applying Cauchy–Schwarz's inequality, we obtain

$$|(I)| \leq \sqrt{\tilde{C}_{1n}\tilde{C}_{2n}},$$

where

$$\tilde{C}_{1n} := \sum_{i=1}^{n} \left(\sqrt{\tau_m^{i-1}} \left(\sqrt{\frac{Z_{i-1}+1}{\tau_m^{i-1}}} - \sqrt{W} \right)^2 \right),$$

$$\tilde{C}_{2n} := \sum_{i=1}^{n} \eta_{ni}^2 \sqrt{\tau_m^{i-1}}$$

and

$$|(I)| = o_p\left(\sqrt{\tau_m^n}\right) \quad \text{as} \quad n \to \infty. \qquad [5.38]$$

Combining [5.38] and [5.33], we obtain

$$\lim_{n\to\infty} \frac{1}{\sqrt{A_n}}(I) \to 0 \ \text{ a.s.} \qquad [5.39]$$

The claim of the theorem follows from [5.35], [5.37], and [5.39]. ∎

We can also show that the estimator $\hat{m}_n$ is strongly consistent in all three criticality modes. Indeed, we have that

$$\hat{m}_n - m = \frac{1}{A_n} \sum_{i=1}^{n} \frac{(Z_i - m\varepsilon(Z_{i-1}))\varepsilon(Z_{i-1})}{1+\varepsilon(Z_{i-1})},$$

where $\lim_{n\to\infty} A_n = \infty$ and

$$\sum_{i=1}^{n} \frac{Z_i - m\varepsilon(Z_{i-1})}{1+\varepsilon(Z_{i-1})}$$

is a martingale. Hence, using again theorem A1.3 in Appendix 1 (or as in the proof of theorem 5.6, part b), we obtain that $\lim_{n\to\infty} \hat{m}_n = m$ a.s.

The proposed CWLS estimator requires that the mean function $\varepsilon(k)$ is known. This condition might not be valid for some CB processes. For example, in the case of GWI processes with unknown immigration mean. If $\varepsilon(k)$ is unknown, then $\varepsilon(Z_{i-1})$ in [5.24] can be replaced by $\phi_{i-1}(Z_{i-1})$. The resulting modified CWLS estimator is given by

$$\hat{\hat{m}}_n := \frac{1}{\tilde{A}_n} \sum_{i=1}^{n} \frac{Z_i \phi_{i-1}(Z_{i-1})}{\phi_{i-1}(Z_{i-1}) + 1}, \qquad \tilde{A}_n := \sum_{i=1}^{n} \frac{\phi_{i-1}^2(Z_{i-1})}{\phi_{i-1}(Z_{i-1}) + 1}.$$

Following arguments like in the proof of theorem 5.7, it can be proved that in the supercritical case the limit distribution of $\sqrt{\tilde{A}_n}(\hat{\hat{m}}_n - m)$ is $N(0, \sigma^2)$. However, the asymptotic behavior of $\hat{\hat{m}}_n$ in the subcritical and critical cases are not known yet. The results presented here extend those of [WEI 90] for the estimation of the offspring mean in the case of known immigration mean.

5.3. Minimum disparity estimation

A small group of individuals in a population with reproductive capacity influenced by temporary events can provoke outliers in the model. For instance, this may occur due to the presence of a disease with a low prevalence or punctual changes on the environmental conditions. It is well-known that MLEs are very sensitive to outliers. Therefore, there is a need for modifications that will produce estimators attaining robustness properties without loss of efficiency. Recently, in [GON 17] robust estimation procedures for CB processes were developed by making use of the **minimum disparity methodology**. This methodology was introduced in [LIN 94] for discrete models and since then, it was developed rapidly (see [PAR 06, BAS 11] and [KUC 17] for further information). Here, we will outline this concept in the context of CB processes.

Let us assume that the offspring distribution belongs to a general parametric family $\mathcal{F}_\Theta = \{p(\theta) : \theta \in \Theta\}$, where $p(\theta) = \{p_k(\theta)\}_{k\geq 0}$ and Θ is a subset of $\mathbf{R}$, that is $p = p(\theta_0)$ for some offspring parameter $\theta_0 \in \Theta$. For simplicity, we assume that θ is a scalar, although the procedure can be generalized for a vector value parameter. Now, we are in a position to introduce the notion of a **disparity measure**. The MDE for θ_0 provides the best adjustment to the observed sample in terms of the chosen disparity measure.

Let Γ be the set of all probability distributions defined on the non-negative integers, $\mathcal{F}_\Theta$ a parametric family and $G(\cdot)$ a three times differentiable and strictly convex function on $[-1, \infty)$ with $G(0) = 0$.

DEFINITION 5.1.– *The disparity measure ρ_G corresponding to $G(\cdot)$ is defined for any $q \in \Gamma$ and $\theta \in \Theta$ as $\rho_G : \Gamma \times \Theta \to [0, \infty]$, such that*

$$\rho_G(q, \theta) = \sum_{k=0}^{\infty} G(\delta(q, \theta, k))p_k(\theta)$$

where $\delta(q, \theta, k)$ denotes the Pearson residual at k given by

$$\delta(q, \theta, k) = \begin{cases} \dfrac{q_k}{p_k(\theta)} - 1, & \text{if } p_k(\theta) > 0; \\ 0, & \text{otherwise.} \end{cases}$$

In order to simplify the notation, we will drop G in ρ_G. Notice that Pearson residual at k depends on the probability distribution q and on the parameter θ, and that $\delta(q, \theta, k)$ belongs to $[-1, \infty)$. Since $G(\cdot)$ is strictly convex, we have $\rho \geq 0$. Moreover, if $G(\cdot)$ is non-negative and has a unique zero at 0, then $\rho(q, \theta) = 0$ if and only if $q = p(\theta)$.

To define an MDE consider some non-parametric estimators for the offspring distribution. Denote by $\tilde{p}_n = \{\tilde{p}_{k,n}\}_{k \geq 0}$ a generic non-parametric estimator for p based on the sample $\mathcal{X}_n$, say, where n indicates that we observe the data up to the nth generation. It satisfies the conditions $\tilde{p}_{k,n} \geq 0$ for $k \geq 0$, and $\sum_{k=0}^{\infty} \tilde{p}_{k,n} = 1$.

DEFINITION 5.2.– *Given a sample $\mathcal{X}_n$ and a non-parametric estimator $\tilde{p}_n$, based on it, the MDE for θ_0 with respect to the disparity measure ρ is given by*

$$\tilde{\theta}_n^{\rho}(\tilde{p}_n) = \arg\min_{\theta \in \Theta} \rho(\tilde{p}_n, \theta). \quad [5.40]$$

Some interesting cases of non-negative disparity measures are the following.

1) **Likelihood Disparity** for which

$$G(\delta) := (\delta + 1)\log(\delta + 1)$$

is a kind of the Kullback–Leibler divergence. The corresponding MDE is called minimum likelihood disparity estimator (MLDE). Under certain conditions, MLDE and MLE coincide.

2) **Squared Hellinger Distance** ($HD(\tilde{p}_n, \theta)$) for which

$$G(\delta) := \left(\sqrt{\delta + 1} - 1\right)^2.$$

Note that for two probability distributions q and $p(\theta)$, say, we have

$$HD(q, \theta) = \|\sqrt{q} - \sqrt{p(\theta)}\|_2^2,$$

where $\|\cdot\|_2$ is the l_2-norm on Γ and $\sqrt{q} = \{\sqrt{q_k}\}_{k \geq 0}$, for each $q \in \Gamma$. The corresponding MDE is called minimum Hellinger distance estimator.

3) **Negative Exponential Disparity** for which

$$G(\delta) := \exp(-\delta) - 1 \qquad \text{or} \qquad G(\delta) := \exp(-\delta) - 2.$$

The MDE is now called the minimum negative exponential disparity estimator.

The existence and uniqueness of the minimum in [5.40] are studied in [GON 17]. In order to establish the asymptotic normality of the MDEs, we need explicit formulas for the non-parametric estimators. For more details about the particular cases of non-parametric MLEs for p, introduced in theorem 5.1, we refer to [GON 17]. The robustness of the estimators is also studied under possible contamination, i.e. when the offspring distribution is subject to the perturbation

$$p(\alpha, \theta, L) = (1 - \alpha)p(\theta) + \alpha\eta_L,$$

where $\alpha \in (0, 1)$ and η_L is a point mass distribution at a non-negative integer L. This model is known as a mixture model for gross errors at L. In [GON 17], explicit characterizations of the breakdown point for negative exponential disparity and squared Hellinger distance are obtained.

5.4. Bayesian inference

This section is devoted to the development of Bayesian inference in the context of CB processes. We consider the same framework as in section 5.1. That is, we have a CB process with offspring distribution $p = \{p_k\}_{k \geq 0}$, without assuming any restrictions on its support, with mean m and finite variance σ^2. Also the control distribution belong to the power series family [5.1], i.e. for every $k \geq 0$,

$$P(\phi_n(k) = j) = \frac{a_k(j)\theta^j}{A(k, \theta)}, \quad j = 0, 1, \ldots; \theta \in \Theta, \qquad [5.41]$$

where $\Theta = \{\theta > 0 : 0 < A(k,\theta) < \infty$ for all $k = 0, 1, 2, \ldots\}$, with $A(k,\theta) = \sum_{j=0}^{\infty} a_k(j)\theta^j$. Moreover, assume for $k = 0, 1, \ldots$

$$A(k,\theta) = b(\theta)^{\varphi(k)},$$

where $b(\cdot)$ is a differentiable function and $\varphi : \mathbf{N}_0 \to \mathbf{N}_0$. This particular form of $A(k,\theta)$ will allow us to use the conjugate prior theory.

5.4.1. *Estimation based on entire family tree up to nth generation*

Assume that the entire family tree up to the nth generation is observed, i.e. $\{X_{l,j} : j = 1, \ldots, \phi_l(Z_l);\ l = 0, 1, \ldots, n-1\}$, or at least the variables

$$Y_l(k) := \sum_{j=0}^{l} Z_j(k), \qquad Z_j(k) = \sum_{l=1}^{\phi_j(Z_j)} \mathcal{I}\{X_{j,l} = k\},\ k \geq 0.$$

Intuitively, $Y_l(k)$ is the total number of progenitors (up to the l-th generation) with exactly k offspring. Recall the notation

$$\mathcal{Z}_n^* = \{Z_l(k),\ l = 0, 1, \ldots, n-1;\ k = 0, 1, \ldots\}, \quad \Delta_{n-1} = \sum_{l=0}^{n-1} \phi_l(Z_l).$$

According to theorem 5.1, the likelihood based on $\mathcal{Z}_n^*$ is

$$\mathcal{L}(p,\theta \mid \mathcal{Z}_n^*) \propto \theta^{\Delta_{n-1}} b(\theta)^{-\sum_{j=0}^{n-1}\varphi(Z_j)} \prod_{k=0}^{\infty} p_k^{Y_{n-1}(k)}. \qquad [5.42]$$

Since offspring and control distributions are independent, an appropriate conjugate prior $\pi(p,\theta)$ for (p,θ) is

$$\pi(p,\theta) = \pi(p)\pi(\theta), \qquad p \sim \mathrm{DP}(p(0), \alpha), \qquad [5.43]$$

where DP denotes the Dirichlet process (see [FER 73]) with base measure $p(0) = \{p_k(0)\}_{k\geq 0}$ and concentration parameter $\alpha > 0$. Furthermore,

$$\pi(\theta) = \frac{\theta^a b(\theta)^{-d}}{\psi(a,d)}, \qquad \psi(a,d) = \int_\Theta \theta^a b(\theta)^{-d} d\theta, \quad a, d \geq 0. \qquad [5.44]$$

Using [5.42]–[5.44], for the posterior distribution we obtain

$$\pi(p,\theta \mid \mathcal{Z}_n^*) \propto \pi(p \mid \mathcal{Z}_n^*)\pi(\theta \mid \mathcal{Z}_n^*), \qquad [5.45]$$

where $\pi(p \mid \mathcal{Z}_n^*)$ is the distribution of the Dirichlet process

$$p \mid \mathcal{Z}_n^* \sim \text{DP}\left(\frac{\alpha}{\alpha+\Delta_{n-1}}p(0) + \frac{1}{\alpha+\Delta_{n-1}}\sum_{k=0}^{\infty} Y_{n-1}(k)\delta_k,\ \alpha+\Delta_{n-1}\right),$$

where δ_k is the Dirac delta at $k \geq 0$ and

$$\pi(\theta \mid \mathcal{Z}_n^*) = \frac{\theta^{a+\Delta_{n-1}} b(\theta)^{-\left(d+\sum_{j=0}^{n-1}\varphi(Z_j)\right)}}{\psi(a+\Delta_{n-1}, d+\sum_{j=0}^{n-1}\varphi(Z_j))}.$$

From [5.45], using the properties of the Dirichlet process, it is straightforward to obtain that the Bayes estimators for the offspring distribution and θ under the squared error loss are

$$\widehat{p}_k = \frac{\alpha p_k(0) + Y_{n-1}(k)}{\alpha+\Delta_{n-1}} \quad \text{and} \quad \widehat{\theta} = \frac{\psi\left(a+\Delta_{n-1}+1, d+\sum_{j=0}^{n-1}\varphi(Z_j)\right)}{\psi\left(a+\Delta_{n-1}, d+\sum_{j=0}^{n-1}\varphi(Z_j)\right)}.$$

As a result, one obtains that the Bayes estimators for m and μ based on the sample $\mathcal{Z}_n^*$, under squared error loss, are given by

$$\widetilde{m} = \frac{\alpha m^{(0)} + \sum_{j=1}^{n} Z_j}{\alpha+\Delta_{n-1}} \quad \text{and} \quad \widetilde{\mu} = \int_\Theta \mu(\theta)\pi(\theta \mid \mathcal{Z}_n^*)d\theta, \qquad [5.46]$$

where $m^{(0)}$ is the mean of $p(0)$. Finally, if there exists $\varphi > 0$, such that $\lim_{k\to\infty}\varphi(k)/k = \varphi$, then the Bayes estimator of the criticality parameter τ_m under squared error loss is $\widetilde{\tau}_m = \varphi\widetilde{m}\widetilde{\mu}$.

If the control distribution [5.41] is either Binomial or Negative Binomial given in section 5.1, then [5.44] yields a Beta conjugate prior for the parameters q and θ, respectively. In case of Poisson control distribution, an appropriate conjugate class of priors for θ is the Gamma distribution. Let us point out that from a practical standpoint, the choice of the control process to be Poisson, Binomial, Negative Binomial or some other should be a prior specification based on specific knowledge of the population's development.

5.4.2. *MCMC algorithms for incomplete data*

In this section, we assume that only the generation sizes $\mathcal{Z}_n = \{Z_0, \ldots, Z_n\}$ are observable. Given the definition of the process, an expression of the posterior distribution for (p, θ) after observing $\mathcal{Z}_n$ cannot be obtained in the closed form. To approximate it, we will describe a MCMC algorithm based on the Gibbs sampler (see, e.g. [BRO 98a]). To this end, it is necessary to consider the unobservable variables $Z_l(k)$ for $l = 0, 1, \dots, n-1$ as latent variables and form the augmented parameter vector $(p, \theta, \mathcal{Z}_n^*)$. We will approximate the posterior distribution $\pi(p, \theta, \mathcal{Z}_n^* \mid \mathcal{Z}_n)$, say, of $(p, \theta, \mathcal{Z}_n^*)$ after observing $\mathcal{Z}_n$ and then calculate an approximation for its marginal distribution $\pi(p, \theta \mid \mathcal{Z}_n)$.

To use the Gibbs sampler, we need to obtain both the conditional posterior distribution of (p, θ) after observing $\mathcal{Z}_n$ and $\mathcal{Z}_n^*$, denoted by $\pi(p, \theta \mid \mathcal{Z}_n, \mathcal{Z}_n^*)$, and the conditional posterior distribution of $\mathcal{Z}_n^*$ after observing $(p, \theta, \mathcal{Z}_n)$, denoted by $f(\mathcal{Z}_n^* \mid p, \theta, \mathcal{Z}_n)$. Since it is verified that $Z_{l+1} = \sum_{k=0}^{\infty} k Z_l(k)$ for $l = 0, \dots, n-1$, we have that $\pi(p, \theta \mid \mathcal{Z}_n, \mathcal{Z}_n^*)$ is the same as $\pi(p, \theta \mid \mathcal{Z}_n^*)$ given in [5.45]. Now, consider $f(\mathcal{Z}_n^* \mid p, \theta, \mathcal{Z}_n)$. Let $P(\cdot)$ be the conditional probability given offspring distribution p and control distributions governed by θ. According to [5.19]

$$P\big(Z_l(k) = z_l(k),\ l = 0, \dots, n-1, k \geq 0 \,\big|\, Z_0 = z_0, \dots, Z_n = z_n\big) \qquad [5.47]$$

$$= \prod_{l=0}^{n-1} \frac{a_{z_l}(\phi_l^*)\theta^{\phi_l^*}}{P\big(Z_{l+1} = z_{l+1} | Z_l = z_l\big)} \frac{\phi_l^*!}{A(z_l, \theta) \prod_{k=0}^{\infty} z_l(k)!} \prod_{k=0}^{\infty} p_k^{z_l(k)}.$$

This leads to the following computational way to sample from $f(\mathcal{Z}_n^* \mid p, \theta, \mathcal{Z}_n)$. Given the known sample $\{z_0, \dots, z_n\}$ and known values of θ and p, sample a value $\phi_l^*(z_l)$ for every $l = 0, 1, \dots, n-1$ from the distribution of $\phi_l(z_l)$ given by [5.1]. Then, sample a sequence $\{z_l(k)\}_{k \geq 0}$ from the multinomial probabilities

$$\frac{\phi_l^*(z_l)!}{\prod_{k \geq 0} z_l(k)!} \prod_{k=0}^{\infty} p_k^{z_l(k)},$$

for every $l = 0, 1, \dots, n-1$, normalized, such that $z_{l+1} = \sum_{k=0}^{\infty} k z_l(k)$. Note that, although the offspring distribution may have an infinite support, once z_{l+1} is known, only a finite number of coordinates in $\{z_l(k), k \geq 0\}$ are different from zero. In fact, $z_l(k) = 0$ for all $k \geq z_{l+1}$.

Once samples from the distributions $\pi(p, \theta \mid \mathcal{Z}_n, \mathcal{Z}_n^*)$ and $f(\mathcal{Z}_n^* \mid p, \theta, \mathcal{Z}_n)$ are available, the Gibbs sampler algorithm works as follows:

Initialize $l = 0$

Generate $p^{(0)} \sim$ DP$(p(0), \alpha)$

```
Generate θ^(0) from [5.44]
Iterate
  l = l + 1
  Generate Z_n^{*(l)} ~ f(Z_n^* | p^(l-1), θ^(l-1), Z_n)
  Generate (p^(l), θ^(l)) ~ π(p, θ | Z_n^{*(l)})
```

Note that, given $\mathcal{Z}_n$, the maximum number of coordinates of $p^{(l)}$, for all $l \geq 0$, involved in the algorithm is $1 + \max_{1 \leq k \leq n} Z_k$. Hence, in the last step of the algorithm, taking into account [5.45] and the properties of the Dirichlet process, these probabilities are calculated from the Dirichlet distribution.

The sequence $\{(p^{(l)}, \theta^{(l)}, \mathcal{Z}_n^{*(l)})\}_{l \geq 0}$ is an ergodic Markov chain. Its stationary distribution is just the sought-after joint distribution $\pi(p, \theta, \mathcal{Z}_n^* \mid \mathcal{Z}_n)$. Several practical implementation issues must be taken into account for a successful application of the method described above. Common approaches to reaching the equilibrium distribution as well as to reducing the autocorrelation in the sample are choosing a sufficient burn-in period, N, and thinning the output by storing only every Gth value after the burn-in period (G is known as the batch size). Thus, for a run of the sequence $\{(p^{(l)}, \theta^{(l)}, \mathcal{Z}_n^{*(l)})\}_{l \geq 0}$, one chooses $Q + 1$ vectors

$$\{(p^{(N)}, \theta^{(N)}), (p^{(N+G)}, \theta^{(N+G)}), \ldots, (p^{(N+QG)}, \theta^{(N+QG)})\}.$$

If G and N are large enough, then these vectors are approximately independent sampled values of the distribution $\pi(p, \theta | \mathcal{Z}_n)$ (see [TIE 94]). Since they could be affected by the initial state $(p^{(0)}, \theta^{(0)})$, the algorithm is applied T times, resulting in a final sample of length $T(Q + 1)$. In practice, in determining N, G and T, the Gelman–Rubin–Brooks and autocorrelation diagnostics need to be used (see [BRO 98b] and [GEL 92]). From this sample $\pi(p, \theta \mid \mathcal{Z}_n)$ and its marginal distributions, $\pi(p \mid \mathcal{Z}_n)$ and $\pi(\theta \mid \mathcal{Z}_n)$ are estimated, by making use of kernel density estimators (see [BOW 97]). These posterior densities can be used to calculate highest–probability–density (HPD) credible sets for the respective parameters. In general, if $\Psi(p, \theta)$ is a function of the offspring distribution and the control parameter, then

$$\pi(\Psi \mid \mathcal{Z}_n) = \int \pi(\Psi \mid \mathcal{Z}_n, p, \theta)\pi(p, \theta \mid \mathcal{Z}_n) dp d\theta.$$

Using kernel density estimators, as before, one can approximate $\pi(\Psi \mid \mathcal{Z}_n)$ and calculate its HPD sets.

A comprehensive study of the methodology described here, including development of methods for forecasting the size of future generations and simulated examples, can be found in [GON 16a].

5.5. Background and bibliographical notes

Good references for statistical analysis of branching processes are the books: [BAD 90, GUT 91], and the reviews: [DIO 93, YAN 08]. A pioneering paper on statistics of CB processes is [DIO 95]. It studies the so-called multiplicative CB process with $\phi_n(Z_n) = \alpha_n\phi(Z_n)$, where $\{\alpha_n\}_{n\geq 0}$ is a sequence of i.i.d. random variables with values in $\mathbb{N}^+$, $E[\alpha_0] = \alpha$ and $Var[\alpha_0] < \infty$. An estimator of $\theta = m\alpha$ is constructed, which is proved to be asymptotically normal in the supercritical case. The investigations of multiplicative CB processes continued in [KAL 00], where estimators for the offspring mean and variance are considered. In [BER 99] for a CB process with adaptive tracking situations, weighted estimators for the offspring mean and variance are studied. Inference problems for a CB process with deterministic control function were investigated in [GON 04a, GON 05a]. The joint asymptotic properties of the estimators for the offspring mean and variance proposed in the last two papers are established in [INA 16]. It is worth mentioning that the proof of proposition 5.1 was provided by C. Minuesa in her unpublished master's degree thesis (University of Extremadura, Spain). Statistical inference based on partially observed GWI process is given in [RAH 17].

EM algorithms have been successfully used in approximating MLEs when there are missing or incomplete data, although there are only a few articles on this topic in the context of branching processes (see [VEE 08], [GON 12b] [DAS 14], [HAU 14]), and none for models with random control except [GON 16b].

Results obtained in [WEI 90] on weighted conditional least squares estimators in the context of GWI processes were extended in [SRI 07]. Moreover, in [GON 10b] a weighted conditional least squares estimator for the offspring mean matrix for a multitype CB process is considered, proving the strong consistency of the proposed estimator in the supercritical case.

Robust estimation for branching processes is barely developed. For GW processes, the robustness of estimators is addressed in [STO 04] by weighted least trimmed estimation and in [SRI 00] by minimum Hellinger distance estimation. Following this research line in [GON 17], the technique of minimum disparity for branching structure given by CB processes is dealt with.

Only a few papers on Bayesian inference for CB processes exist. Assuming power series offspring distribution and observing both population size and number of progenitors in every generation, the asymptotic normality of Bayes estimators under squared error loss is proved in [MAR 09]. Particular results for the GWI process are also established. Computational methodologies, as MCMC and approximate Bayesian computation (ABC), are explored in [GON 13] for CB processes with deterministic control. Finally, in [GON 16a], Bayesian estimators are obtained based on complete and incomplete data frames for CB processes with random control.

Appendices

Appendix 1

Limit Theorems for Martingales

A1.1. Martingales

DEFINITION A1.1.– *A sequence $\{X_n\}_{n\geq 0}$ of integrable random variables is called a* **martingale** *with respect to an increasing sequence $\{\mathcal{F}_n\}_{n\geq 0}$ of σ-algebras (we also write $\{X_n, \mathcal{F}_n\}_{n\geq 0}$ is a martingale) if X_n is $\mathcal{F}_n$-measurable and $E[X_{n+1} \mid \mathcal{F}_n] = X_n$. If instead $E[X_{n+1} \mid \mathcal{F}_n] \geq X_n$ or $E[X_{n+1} \mid \mathcal{F}_n] \leq X_n$, we talk of a* **submartingale** *or a* **supermartingale**, *respectively.*

If $\{X_n\}_{n\geq 0}$ is a martingale with respect to $\{\mathcal{F}_n\}_{n\geq 0}$ and we define $Y_n := X_n - X_{n-1}$, then clearly $E[Y_{n+1} \mid \mathcal{F}_n] = 0$. The sequence of random variables with this property is called **martingale difference** or **martingale increment**.

A1.2. Basic convergence theorem (Doob's theorem)

THEOREM A1.1.– *Any submartingale $\{X_n, \mathcal{F}_n\}_{n\geq 0}$ satisfying*

$$\sup_n E[|X_n|] < \infty \qquad [A1.1]$$

converges a.s. to a (finite) random variable X_∞. Furthermore, $E[|X_\infty|] < \infty$.

The condition [A1.1] is known as **Doob's condition**. Two very useful corollaries of theorem A1.1 are as follows:

1) If $\{X_n\}_{n\geq 0}$ is a non-negative martingale, i.e. $X_n \geq 0$ a.s., then [A1.1] holds and hence the a.s. convergence is automatic. More generally, any non-negative supermartingale converges a.s. converges.

2) The condition $\sup_n E[X_n^2] < \infty$ implies [A1.1] and hence a.s. convergence.

In connection with (2), it is often useful to note that the increments of an L^2–martingale are orthogonal in L^2, e.g.

$$E[X_n^2] = E[X_0^2] + \sum_{k=0}^{n-1} E\left[(X_{k+1} - X_k)^2\right] = E[X_0^2] + \sum_{k=0}^{n-1} Var[X_{k+1} - X_k],$$

and thus

$$\sum_{k=0}^{\infty} Var[X_{k+1} - X_k] < \infty \qquad [A1.2]$$

is a sufficient condition for a.s. convergence.

THEOREM A1.2.– *Any martingale satisfying* [A1.2] *converges a.s.*

The proof of Doob's theorem and its corollaries can be found in [WIL 91, p. 109].

A1.3. A strong law of large numbers for martingales

THEOREM A1.3.– *Let $\{X_n\}_{n\geq 0}$ be a sequence of random variables on a probability space $(\Omega, \mathcal{F}, P)$ and let $\{\mathcal{F}_n\}_{n\geq 0}$ be a sequence of sub-σ-algebras of $\mathcal{F}$. Let $\{U_n\}_{n\geq 1}$ be a sequence of non–decreasing positive random variables, such that U_n is $\mathcal{F}_{n-1}$–measurable for every n. If $1 \leq p \leq 2$, then*

$$\sum_{i=1}^{\infty} \frac{X_i}{U_i} \quad \textit{converges a.s.} \qquad [A1.3]$$

$$\textit{on the set} \left\{ \sum_{i=1}^{\infty} \frac{E[|X_i|^p \mid \mathcal{F}_{i-1}]}{U_i^p} < \infty \right\}, \textit{ and}$$

$$\lim_{n\to\infty} \frac{1}{U_n} \sum_{i=1}^{n} X_i = 0 \quad a.s. \qquad [A1.4]$$

on the set $\left\{ \lim_{n\to\infty} U_n = \infty, \sum_{i=1}^{\infty} \frac{E[|X_i|^p \mid \mathcal{F}_{i-1}]}{U_i^p} < \infty \right\}$.

If $2 < p < \infty$ *then* [A1.3] *and* [A1.4] *are verified on the set*

$$\left\{ \sum_{i=1}^{\infty} \frac{1}{U_i} < \infty, \sum_{i=1}^{\infty} \frac{E[|X_i|^p \mid \mathcal{F}_{i-1}]}{U_i^{1+p/2}} < \infty \right\}.$$

The proof can be found in [HAL 80, p. 36].

A1.4. A central limit theorem for martingales

If $\{Y_n\}_{n\geq 0}$ is a sequence of random variables on a probability space $(\Omega, \mathcal{F}, P)$ converging in distribution to a random variable Y, it is said that the convergence is stable if for all continuity points ξ of Y and all events $E \in \mathcal{F}$,

$$\lim_{n\to\infty} P(\{Y_n \leq \xi\} \cap E) = Q_\xi(E)$$

exists, and $Q_\xi(E) \to P(E)$ as $\xi \to \infty$ (clearly Q_ξ, if it exists, is a probability measure on $(\Omega, \mathcal{F})$). We designate the convergence by writing $Y_n \xrightarrow{d} Y$ (stably).

Let $\{S_{ni}, \mathcal{F}_{ni}, 1 \leq i \leq k_n\}$ be a zero-mean, square-integrable martingale for each $n \geq 1$, and let $X_{ni} = S_{ni} - S_{n,i-1}$, $1 \leq i \leq k_n$ ($S_{n0} = 0$) denote the martingale differences (suppose that $k_n \to \infty$ as $n \to \infty$). We shall call the double sequence $\{S_{ni}, \mathcal{F}_{ni}, 1 \leq i \leq k_n\}_{n\geq 1}$ a martingale array.

THEOREM A1.4.– *Let* $\{S_{ni}, \mathcal{F}_{ni}, 1 \leq i \leq k_n, n \geq 1\}$ *be a zero-mean, square-integrable martingale array with differences* $X_{ni} = S_{ni} - S_{n,i-1}$. *Let* η^2 *be an a.s. finite random variable. Suppose that*

1) $V_{nk_n}^2 = \sum_{i=1}^{k_n} E[X_{ni}^2 | \mathcal{F}_{n,i-1}] \xrightarrow{P} \eta^2$ *as* $n \to \infty$,

2) *for every* $\epsilon > 0$,

$$\sum_{i=1}^{k_n} E[X_{ni}^2 \mathcal{I}\{|X_{ni}| > \epsilon\} | \mathcal{F}_{n,i-1}] \xrightarrow{P} 0 \text{ as } n \to \infty$$

(the conditional Lindeberg condition),

3) $\mathcal{F}_{ni} \subseteq \mathcal{F}_{n+1,i}$, $1 \leq i \leq k_n$, $n \geq 1$

then

$$S_{nk_n} = \sum_{i=1}^{k_n} X_{ni} \xrightarrow{d} Z(stably) \quad as\ n \to \infty$$

where the random variable Z *has characteristic function* $E[\exp(-\eta^2 t^2/2)]$.

The proof can be found in [HAL 80, p. 58].

Appendix 2

Some Classical Theorems

A2.1. Slowly and regularly varying functions

DEFINITION A2.1.– *Slowly varying function. A measurable, real-valued function $L(x)$, defined on $[0, \infty)$, is called slowly varying (at infinity) (s.v.f.) if it is positive and if for each $\lambda > 0$*

$$\lim_{x \to \infty} \frac{L(\lambda x)}{L(x)} = 1.$$

A measurable, real-valued function $R(x)$ is called regularly varying (at infinity) with exponent α, $\alpha \in (-\infty, \infty)$, if it is of the form

$$R(x) = x^{\alpha} L(x),$$

with L slowly varying at infinity.

A2.2. Marcinkiewicz–Zygmund's inequality

LEMMA A2.1.– *If $\{X_n\}_{n \geq 1}$ is a sequence of i.i.d. random variables with $E[X_1] = 0$ and $E[|X_1|^p] < \infty$, $p \geq 2$, then $E[|\sum_{j=1}^{n} X_j|^p] = O(n^{p/2})$.*

The proof of this lemma can be found in [CHO 97, p. 387].

A2.3. Cèsaro's lemma

LEMMA A2.2.– *Let $\{a_n\}_{n \geq 0}$ be a sequence of real positive numbers such that $a_n \to \infty$ as $n \to \infty$ and $\{x_n\}_{n \geq 0}$ be a sequence of real numbers such that $x_n \to x$ as $n \to \infty$. Then*

$$\frac{1}{a_n} \sum_{k=1}^{n} (a_k - a_{k-1}) x_k \to x, \text{ as } n \to \infty.$$

The proof can be found in [WIL 91, p. 117].

A2.4. Slutsky's theorem

THEOREM A2.1.– *Let* $\{X_n\}_{n\geq 1}$ *and* $\{Y_n\}_{n\geq 1}$ *be sequences of random variables defined on the same probability space such that* $X_n - Y_n \xrightarrow{P} 0$ *and* $Y_n \xrightarrow{d} Z$, *as* $n \to \infty$, *then* $X_n \xrightarrow{d} Z$, *as* $n \to \infty$.

COROLLARY A2.1.– *The following statements are true:*

1) Let $\{X_n\}_{n\geq 1}$, $\{Y_n\}_{n\geq 1}$ *and* $\{Z_n\}_{n\geq 1}$ *be sequences of random variables defined on the same probability space such that* $X_n \xrightarrow{d} X$, $Y_n \xrightarrow{P} a$ *and* $Z_n \xrightarrow{P} b$, *as* $n \to \infty$, *with* a *and* b *being finite constants, then* $X_nY_n + Z_n \xrightarrow{d} aX + b$, *as* $n \to \infty$.

2) Let a, b, a_n, b_n, $n = 1, 2, \ldots$ *be finite constants such that* $a_n \to a$, $b_n \to b$, *as* $n \to \infty$ *and let* $\{X_n\}_{n\geq 1}$ *be a sequence of random variables such that* $X_n \xrightarrow{d} X$, *as* $n \to \infty$, *then* $a_nX_n + b_n \xrightarrow{d} aX + b$, *as* $n \to \infty$.

The proofs can be found in [CHO 97, p. 272].

A2.5. Doeblin–Anscombe's theorem

THEOREM A2.2.– *Let* $\{X_n\}_{n\geq 1}$ *be a sequence of independent random variables with* $E[X_n] = 0$ *and* $E[X_n^2] = 1$, *and* $\{t_n\}_{n\geq 1}$ *be a sequence of positive integer-valued random variables with* $t_n/b_n \xrightarrow{P} c$, *where* $\{b_n\}_{n\geq 1}$ *and* c *are finite positive constants such that* $b_n \to \infty$ *as* $n \to \infty$. *If* $\frac{1}{\sqrt{n}}\sum_{j=1}^{n} X_j \xrightarrow{d} Z$, *as* $n \to \infty$, *with the random variable* Z *following a standard Normal distribution,* $N(0,1)$, *then, as* $n \to \infty$,

$$\frac{1}{\sqrt{t_n}}\sum_{j=1}^{t_n} X_j \xrightarrow{d} Z.$$

The proof can be found in [CHO 97, p. 317].

A2.6. Conditional Borel–Cantelli lemma

Let $\{X_n\}_{n\geq 0}$ be a sequence of non-negative random variables defined on $(\Omega, \mathcal{F}, P)$ and let $\{\mathcal{F}_n\}_{n\geq 0}$ be an increasing sequence of sub-σ-algebras of $\mathcal{F}$.

LEMMA A2.3.– *It is verified that*

$$\sum_{n=1}^{\infty} X_n < \infty \text{ a.s. on } \left\{ \sum_{n=1}^{\infty} E\left[X_n | \mathcal{F}_n\right] < \infty \right\}.$$

Moreover, if X_n *is* $\mathcal{F}_{n+1}$*–measurable for all* n*, then*

$$\left\{ \sum_{n=1}^{\infty} X_n < \infty \right\} = \left\{ \sum_{n=1}^{\infty} E\left[X_n | \mathcal{F}_n\right] < \infty \right\} \quad a.s.$$

Let $\{A_n\}_{n\geq 0}$ be a sequence of events on $\mathcal{F}$ and $X_n = \mathcal{I}A_n$. Then

$$\left\{ \sum_{n=1}^{\infty} X_n < \infty \right\} = \liminf_{n\to\infty} A_n^c$$

$$\left\{ \sum_{n=1}^{\infty} X_n = \infty \right\} = \limsup_{n\to\infty} A_n$$

LEMMA A2.4.– *It is verified that*

$$\left\{ \sum_{n=1}^{\infty} P(A_n | \mathcal{F}_n) < \infty \right\} \subseteq \liminf_{n\to\infty} A_n^c \quad a.s.,$$

or equivalently,

$$\limsup_{n\to\infty} A_n \subseteq \left\{ \sum_{n=1}^{\infty} P(A_n | \mathcal{F}_n) = \infty \right\} \quad a.s.$$

Moreover, if $A_n \in \mathcal{F}_{n+1}$ *for all* n*, then*

$$\limsup_{n\to\infty} A_n = \left\{ \sum_{n=1}^{\infty} P(A_n | \mathcal{F}_n) = \infty \right\} \quad a.s.$$

The proofs of these results can be found in [LOÈ 17].

A2.7. A functional limit theorem

Let $\{\xi_n\}_{n\geq 1}$ be a sequence of i.i.d. random variables defined on a probability space $(\Omega, \mathcal{F}, P)$ with $E[\xi_1] = 0$ and $0 < E[\xi_1^2] = \sigma^2 < \infty$,

$i = 1, 2, \ldots$. Consider the partial sums $S_n = \sum_{i=1}^{n} \xi_i$, $n = 1, 2, \ldots$ and define for $0 \leq t \leq 1$

$$X_n(t, \omega) = \frac{S_{[nt]}(\omega)}{\sqrt{\sigma^2 n}}, \quad \omega \in \Omega,$$

where $[x]$ denotes the integer part of x. Let, for each $n \geq 1$, ν_n be a non-negative random variable on $(\Omega, \mathcal{F}, P)$ and consider

$$Y_n(t, \omega) = \begin{cases} X_{\nu_n(\omega)}(t, \omega) & \text{if } \nu_n(\omega) > 0 \\ 0 & \text{otherwise.} \end{cases}$$

THEOREM A2.3.– *Suppose there exist constants* $\{a_n\}_{n \geq 0}$ *such that*

1) $a_n \to \infty$ *as* $n \to \infty$;

2) $\nu_n / a_n \xrightarrow{P} \widetilde{W}$ *as* $n \to \infty$, *where* $\widetilde{W}$ *is a non-negative random variable.*

Suppose also that $\mathcal{D} := \{\omega\colon \widetilde{W}(\omega) > 0\}$ *is such that* $P(\mathcal{D}) > 0$. *Denote* $P_{\mathcal{D}}(\cdot) := P(\cdot | \mathcal{D})$. *If* P^* *is a probability measure dominated by* $P_{\mathcal{D}}$, *then for* $0 \leq t \leq 1$

$$P^*(\omega\colon Y_n(t, \omega) \leq x) \to P^*(\omega\colon B(t, \omega) \leq x) \quad as \quad n \to \infty,$$

where $\{B(t)\}_{0 \leq t \leq 1}$ *is the standard Wiener process.*

The proof can be found in [DIO 74].

A2.8. Continuous theorem for Laplace transforms

DEFINITION A2.2.– *Let* F *be a proper or defective probability distribution concentrated on* $[0, \infty]$. *The Laplace transform* $\phi(\lambda)$ *of* F *is defined for* $\lambda > 0$ *by*

$$\phi(\lambda) := \int_0^\infty e^{-\lambda x} F(dx).$$

It is understood that the interval of integration is closed. We say "the Laplace transform of the random variable X" meaning the transform of its distribution.

THEOREM A2.4.– *For* $n = 1, 2, \ldots$, *let* F_n *be a probability distribution on* $[0, \infty]$ *with Laplace transform* ϕ_n.

1) If $\lim_{n\to\infty} F_n = F$ *where* F *is a possibly defective probability distribution with Laplace transform* ϕ*, then for* $\lambda > 0$*,*

$$\lim_{n\to\infty} \phi_n(\lambda) = \phi(\lambda).$$

2) Conversely, if the sequence $\{\phi_n(\lambda)\}_{n\geq 1}$ *converges for each* $\lambda > 0$ *to a limit* $\phi(\lambda)$*, then* ϕ *is the Laplace transform of a possibly defective distribution* F*, and* $\lim_{n\to\infty} F_n = F$*.*

The limit F *is not defective if and only if* $\lim_{\lambda\to 0} \phi(\lambda) = 1$*.*

The proof can be found in [FEL 71, p. 431].

A2.9. Continuous mapping theorem

THEOREM A2.5.– *Let* S *and* T *be two metric spaces, and* $X, X_1, X_2, \cdots$ *be random elements in* S *with* $X_n \xrightarrow{d} X$*, as* $n \to \infty$*. Consider some measure mappings* $h, h_1, h_2, \cdots : S \to T$ *and a measure set* $C \subset S$ *with* $X \in C$ *a.s. such that* $h_n(s_n) \to h(s)$ *as* $s_n \to s \in C$*. Then* $h_n(X_n) \xrightarrow{d} h(X)$*, as* $n \to \infty$*.*

The proof can be found in [KAL 97, p. 54].

Appendix 3

Auxiliaries

A3.1. A Taylor expansion

LEMMA A3.1.– *For any* $0 \leq k \leq 2$ *and* $x \geq -1$*, it is verified that*

$$(1 + x)^k = 1 + kx + O(x^2).$$

We refer the reader to [KLE 89a] for the proof of this result.

A3.2. Results on stochastic difference equations

Define a discrete time stochastic process $\{X_n\}_{n\geq 0}$ recursively by the following equation,

$$X_{n+1} = X_n + g(X_n)(1 + \eta_{n+1}), \qquad X_0 > 0, \quad E[X_0^2] < \infty,$$

where $g(t) = o(t)$, as $t \to \infty$, is a strictly positive function and $\{\eta_n, \mathcal{F}_n\}_{n>0}$ is a zero-mean, square-integrable martingale difference sequence. Here, $\{\mathcal{F}_n\}_{n\geq 0}$ is a non-decreasing sequence of σ-algebras, such that X_n is $\mathcal{F}_n$-measurable. Assume that there exists a measurable function $\varphi^2(\cdot)$ such that

$$\varphi^2(X_n) := E[\eta_{n+1}^2 | \mathcal{F}_n] \qquad \text{a.s.},$$

i.e. the conditional second moment of η_{n+1} depends only on the present state of the process $\{X_n\}_{n\geq 0}$. Also let for some $\delta > 0$

$$E[|\eta_{n+1}|^{2+\delta} | \mathcal{F}_n] = O\left(\varphi^{2+\delta}(X_n)\right), \qquad n \to \infty.$$

Next, we introduce the auxiliary function

$$G(t) := \int_1^t \frac{1}{g(s)}\, ds, \quad t > 0.$$

Denote by G^{-1} the inverse of G. Define for $t > 0$

$$\psi(t) := \int_1^t \frac{\varphi^2(s)}{g(s)}\, ds \quad \text{and} \quad \widehat{\psi}(t) := \psi\left(G^{-1}(t)\right) := \int_0^t \varphi^2\left(G^{-1}(s)\right) ds.$$

Also, let $\{a_n\}_{n\geq 0}$ be the sequence given by the recursive equation

$$a_{n+1} = a_n + g(a_n)\,, \quad a_0 = 1.$$

Let us formulate some assumptions as follows:

(A.1) $g:\ \mathbf{R} \to \mathbf{R}$ is positive and twice continuously differentiable for $t > 0$ and $g(t) = 0$ on $t < 0$. $g(\cdot)$ and $g'(\cdot)$ are both ultimately concave or convex and $g(t) = o(t)$ as $t \to \infty$.

(A.2) $\varphi^2 : \mathbf{R}^+ \to \mathbf{R}^+$ is positive and continuously differentiable. Additionally, if $\hat{\varphi}^2(t) := \varphi^2(G^{-1}(t))$, then

$$\int_1^\infty \frac{\widehat{\varphi}^2(t)}{t^2}\, dt < \infty$$

and $\widehat{\varphi}^2(t)$ is ultimately concave or convex.

(A.3) If $\lim_{x\to\infty} \widehat{\psi}(x) = \infty$, we suppose that $|g' \circ \psi^{-1}(t)|$ is ultimately convex, which is equivalent to $\lambda(t) := |(g''(t)g(t))/\varphi^2(t)|$ being ultimately decreasing, because $\lambda \circ \psi^{-1}(t) = |(g' \circ \psi^{-1})'(t)|$; whereas if $\lim_{x\to\infty}, \widehat{\psi}(x) < \infty$ we assume that $|g' \circ G^{-1}(t)|$ is ultimately convex or, equivalently, that $|g''(t)g(t)|$ is ultimately decreasing

THEOREM A3.1.– *Assume the conditions [A.1]–[A.3] hold and $g'(t)\psi^{1/2}(t) \to 0$ as $t \to \infty$.*

(a) If $\lim_{x\to\infty} \widehat{\psi}(x) < \infty$, then on $\{X_n \to \infty\}$, as $n \to \infty$,

$$\lim_{n\to\infty} \frac{X_n}{a_n} = 1 \quad a.s. \quad and \quad \lim_{n\to\infty} \frac{X_n - a_n}{g(a_n)} \quad converges\ a.s.$$

(b) If $\lim_{x\to\infty} \widehat{\psi}(x) = \infty$, then on $\{X_n \to \infty\}$, as $n \to \infty$,

$$\frac{X_n}{a_n} \xrightarrow{P} 1 \quad and \quad \frac{1}{\sqrt{\widehat{\psi}(n)}} \frac{X_n - a_n}{g(a_n)} \xrightarrow{d} Z,$$

where the random variable Z follows a standard Normal distribution, $N(0,1)$.

The proof of this theorem can be found in [KEL 87].

Let $\{\mathcal{F}_n\}_{n\geq 0}$ be an increasing sequence of σ-algebras on some probability space and let $\{X_n\}_{n\geq 0}$ be a sequence of real-valued random variables that satisfy the following assumptions:

(B.1) $X_n \geq 0$, X_n is $\mathcal{F}_n$-measurable.

(B.2) There are functions $g(x)$ and $\varphi^2(x)$ for $x \geq 0$, and random variables $\xi_1, \xi_2, \ldots$, such that a.s.

$$\begin{aligned} X_{n+1} &= X_n + g(X_n) + \xi_{n+1}, \\ E[\xi_{n+1} \mid \mathcal{F}_n] &= 0, \\ E[\xi_{n+1}^2 \mid \mathcal{F}_n] &= \varphi^2(X_n). \end{aligned}$$

(B.3) There is a $C > 0$ and $\delta > 2$, such that for all $n \geq 0$ a.s.

$$E\left[|\xi_{n+1}|^\delta \mid \mathcal{F}_n\right] \leq C(\varphi^2(X_n))^{\delta/2}.$$

Let $g(x) > 0$ for $x > 0$ and define, as in the previous result,

$$G(x) := \int_1^x \frac{1}{g(y)}\, dy.$$

(B.4) $g(x)$ is strictly positive and differentiable for $x > 0$. Furthermore, there is a $\lambda \geq 0$, such that $G^\mu(x)$ is ultimately convex for $\mu \notin [0, \lambda]$ and ultimately concave for $\mu \in (0, \lambda)$.

The following limit theorem is proved in [KER 92].

THEOREM A3.2.– *Assume the conditions [B.1]–[B.4] hold and $P(X_n \to \infty) > 0$. If for some $\beta > 0$ with $\beta\lambda < 2$,*

$$\lim_{x\to\infty} \frac{\varphi^2(x)}{g^2(x)G(x)} = \beta,$$

then for any $t \geq 0$,

$$\lim_{n\to\infty} P\left(\frac{G(X_n)}{n} \leq t \mid X_n \to \infty\right) = \Gamma_{a,b}(t),$$

where $\Gamma_{a,b}(\cdot)$ denotes the Gamma c.d.f. with parameters $a = (2/\beta) - \lambda + 1$ and $b = \beta/2$.

Appendix 4

Simulated Data for the Example in Chapter 5

n	Z_n	$\phi_n(Z_n)$	$Z_n(0)$	$Z_n(1)$	$Z_n(2)$	$Z_n(3)$
0	1	1	0	1	0	0
1	1	1	0	1	0	0
2	1	1	0	0	1	0
3	2	1	0	0	1	0
4	2	2	0	0	1	1
5	5	5	0	2	2	1
6	9	6	1	2	2	1
7	9	7	2	2	1	2
8	10	8	0	3	1	4
9	17	14	1	8	3	2
10	20	14	0	8	2	4
11	24	17	2	2	6	7
12	35	25	1	6	8	10
13	52	39	4	11	14	10
14	69	48	10	15	13	10
15	71	38	8	9	14	7
16	58	36	1	5	17	13
17	78	51	5	13	15	18
18	97	61	7	28	13	13
19	93	61	5	22	22	12
20	102	64	6	15	20	23
21	124	72	7	14	25	26
22	142	76	5	24	32	15
23	133	73	9	21	22	21
24	128	81	9	16	33	23
25	151	86	8	21	34	23
26	158	83	7	19	34	23
27	156	94	11	26	32	25
28	165	94	11	29	24	30
29	167	107	10	27	37	33
30	200	.	.	.	.	.

z_{30} $\overline{z}_{30}$ z^*_{30}

Bibliography

[ADK 95] ADKE S.R., GADAG V.G., "A new class of branching processes", in HEYDE C.C. (ed.), *Branching Processes. Proceedings of the First World Congress*, vol. 99 of *'Lecture Notes in Statistics'*, Springer, New York, pp. 90–105, 1995.

[ALL 15] ALLEN L.S., *Stochastic Populations and Epidemic Models. Persistence and Extinction*, vol. 1, part 3 of *'Mathematical Biosciences Institute Lecture Series'*, Springer, New York, 2015.

[ASM 83] ASMUSSEN S., HERING H., *Branching Processes*, Birkhäuser, Boston, 1983.

[ATH 72] ATHREYA K.B., NEY P.E., *Branching Processes*, vol. 196 of *'Die Grundlehren der Mathematischen Wissenschaften'*, Springer, Berlin, 1972.

[BAD 90] BADALBAEV I.S., MUKHITDINOV A., *Statistical Problems in Multitype Branching Processes* (in Russian), FAN, Tashkent, 1990.

[BAG 86] BAGLEY J.H., "On the almost sure convergence of controlled branching processes", *Journal of Applied Probability*, vol. 23, pp. 827–831, 1986.

[BAS 11] BASU A., SHIOYA H., PARK C., *Statistical Inference: The Minimum Distance Approach*, Chapman and Hall/CRC, Boca Raton, 2011.

[BER 99] BERCU B., "Weighted estimation and tracking for Bienaymé–Galton–Watson processes with adaptive control", *Statistics & Probability Letters*, vol. 42, pp. 415–421, 1999.

[BIL 95] BILLINGSLEY P., *Probability and Measure, 3rd ed.*, Wiley, New York, 1995.

[BOR 88] BOROVKOV K.A., "A method of proving limit theorems for branching processes", *Theory of Probability and Its Applications*, vol. 33, pp. 115–123, 1988, in Russian. English translation: vol. 33, pp. 105–113, 1989.

[BOW 97] BOWMAN A., AZZALINI A., *Applied Smoothing Techniques for Data Analysis: The Kernel Approach with S-Plus Illustrations*, Oxford Univiversity Press, Oxford, 1997.

[BRO 98a] BROOKS S., "Markov Chain Monte Carlo method and its application", *Journal of the Royal Statistical Society: Series D. The Statistician*, vol. 47, pp. 69–100, 1998.

[BRO 98b] BROOKS S., GELMAN A., "General methods for monitoring convergence of iterative simulations", *Journal of Computational and Graphical Statistics*, vol. 7, pp. 434–455, 1998.

[BRU 78] BRUSS F.T., "Branching processes with random absorbing processes", *Journal of Applied Probability*, vol. 15, pp. 54–64, 1978.

[BRU 80] BRUSS F.T., "A counterpart of the Borel–Cantelli lemma", *Journal of Applied Probability*, vol. 17, pp. 1094–1101, 1980.

[BRU 84] BRUSS F.T., "A note on extinction criteria for bisexual Galton–Watson processes", *Journal of Applied Probability*, vol. 21, pp. 915–919, 1984.

[BRU 15] BRUSS F.T., DUERINCKX M., "Resource dependent branching processes and the envelope of societies", *The Annals of Applied Probability*, vol. 25, pp. 324–372, 2015.

[CHO 97] CHOW Y.S., TEICHER H., *Probability Theory. Independence, Interchangeability, Martingales*, 3rd ed., Springer, New York, 1997.

[CHU 67] CHUNG K.L., *Markov Chains with Stationary Transition Probabilities*, 2nd ed., Springer, New York, 1967.

[DAL 68] DALEY D., "Extinction condition for certain bisexual Galton-Watson branching process", *Zeitschrift für Wahrscheinlichkeitstheorie und verwandte Gebiete*, vol. 9, pp. 315–322, 1968.

[DAS 14] DASKALOVA N., "Algorithm for estimation of the offspring probabilities in some branching models", in MLADENOV V.M., IVANOV P. (eds), *Nonlinear Dynamics of Electronic Systems*, Springer, Berlin, 2014.

[DIO 74] DION J.P., "Estimation of the mean and the initial probabilities of a branching process", *Journal of Applied Probability*, vol. 11, pp. 687–694, 1974.

[DIO 93] DION J.P., "Statistical inference for discrete time branching processes", in OBRETENOV A., STEFANOV V.T. (eds), *Lecture Notes of 7th Int. Summer School on Probab. Theory and Math. Statistics*, Science Culture Technology, Singapore, 1993.

[DIO 95] DION J.P., ESSEBBAR B., "On the statistics of controlled branching processes", in HEYDE C.C. (ed.), *Branching Processes. Proceedings of the First World Congress*, vol. 99 of *'Lecture Notes in Statistics'*, Springer, New York, pp. 14–21, 1995.

[DRM 06] DRMOTA M., LOUCHARD G., YANEV N.M., "Analysis of a recurrence related to critical non-homogeneous branching processes", *Stochastic Analysis and Applications*, vol. 24, pp. 37–59, 2006.

[DYA 96] DYAKONOVA E.E., "Close-to-critical branching processes with migration", *Theory of Probability and Its Applications*, vol. 41, pp. 186–192, 1996, in Russian. English translation: vol. 41, pp. 151–156, 1997.

[FEL 51] FELLER W., "Diffusion processes in genetics", *Proceedings of the Second Berkeley Symposium on Mathematical Statistics and Probability*, University of California Press, Berkeley, pp. 227–246, 1951.

[FEL 66] FELLER W., *An Introduction to Probability Theory and its Application*, 2nd ed., Volume 1, Wiley, New York, 1966.

[FEL 71] FELLER W., *An Introduction to Probability Theory and its Applications*, Volume 2, Wiley, New York, 1971.

[FER 73] FERGUSON T.S., "A Bayesian analysis of some nonparametric problems", *The Annals of Statistics*, vol. 1, pp. 209–230, 1973.

[FOS 71] FOSTER J.H., "A limit theorem for a branching process with state-dependent immigration", *The Annals of Mathematical Statistics*, vol. 42, pp. 1773–1776, 1971.

[GEL 92] GELMAN A., RUBIN D., "Inference from iterative simulation using multiple sequences", *Statistical Sciences*, vol. 7, pp. 457–511, 1992.

[GON 02] GONZÁLEZ M., MOLINA M., DEL PUERTO I., "On the class of controlled branching processes with random control functions", *Journal of Applied Probability*, vol. 39, pp. 804–815, 2002.

[GON 03] GONZÁLEZ M., MOLINA M., DEL PUERTO I., "On the geometric growth in controlled branching process with random control function", *Journal of Applied Probability*, vol. 40, pp. 995–1006, 2003.

[GON 04a] GONZÁLEZ M., MARTÍNEZ R., DEL PUERTO I., "Nonparametric estimation of the offspring distribution and mean for a controlled branching process", *TEST*, vol. 13, pp. 465–479, 2004.

[GON 04b] GONZÁLEZ M., MOLINA M., DEL PUERTO I., "Limiting distribution for subcritical controlled branching process with random control function", *Statistics and Probability Letters*, vol. 67, pp. 277–284, 2004.

[GON 04c] GONZÁLEZ M., MOLINA M., DEL PUERTO I., "Recent results for supercritical branching processes with random control functions", *Pliska Studia Mathematica Bulgarica*, vol. 16, pp. 43–54, 2004.

[GON 04d] GONZÁLEZ M., MOLINA M., DEL PUERTO I., "On the controlled Galton–Watson process with random control function", *Journal of Mathematical Sciences*, vol. 121, pp. 2629–2635, 2004.

[GON 05a] GONZÁLEZ M., MARTÍNEZ R., DEL PUERTO I., "Estimation of the variance for a controlled branching process", *TEST*, vol. 14, pp. 199–213, 2005.

[GON 05b] GONZÁLEZ M., MARTÍNEZ R., MOTA M., "On the geometric growth in a class of homogeneous multitype Markov chains", *Journal of Applied Probability*, vol. 42, pp. 1015–1030, 2005.

[GON 05c] GONZÁLEZ M., MARTÍNEZ R., MOTA M., "On the unlimited growth of a class of homogeneous multitype Markov chains", *Bernoulli*, vol. 11, pp. 559–570, 2005.

[GON 05d] GONZÁLEZ M., MOLINA M., DEL PUERTO I., "Asymptotic behavior of critical controlled branching processes with random control functions", *Journal of Applied Probability*, vol. 42, pp. 463–477, 2005.

[GON 05e] GONZÁLEZ M., MOLINA M., DEL PUERTO I., "On the L^2-convergence of controlled branching process with random control function", *Bernoulli*, vol. 11, pp. 37–46, 2005.

[GON 06a] GONZÁLEZ M., MARTÍNEZ R., MOTA M., "Rates of growth in a class of homogeneous multidimensional Markov chains", *Journal of Applied Probability*, vol. 43, pp. 159–174, 2006.

[GON 06b] GONZÁLEZ M., MOLINA M., DEL PUERTO I., "Geometric growth for stochastic difference equations with application to branching populations", *Bernoulli*, vol. 12, pp. 931–942, 2006.

[GON 06c] GONZÁLEZ M., MOLINA M., DEL PUERTO I., "Some limit results on controlled branching processes with random control function", *Journal of Mathematical Sciences*, vol. 138, pp. 5396–5404, 2006.

[GON 07] GONZÁLEZ M., MARTÍNEZ R., MOTA M., "Controlled multitype branching models: geometric growth", *Pliska Studia Mathematica Bulgarica*, vol. 18, pp. 103–110, 2007.

[GON 10a] GONZÁLEZ M., DEL PUERTO I., "Branching processes: genealogy and evolution", *Boletín de Estadística e Investigación Operativa*, vol. 26, pp. 107–123, 2010.

[GON 10b] GONZÁLEZ M., DEL PUERTO I., "Weighted conditional least squares estimation in controlled multitype branching processes", in GONZÁLEZ M., DEL PUERTO I., MARTÍNEZ R. *et al.*, (eds), *Workshop on Branching Processes and Their Applications*, vol. 197 of *Lecture Notes in Statistics*, Springer, Berlin, pp. 147–155, 2010.

[GON 12a] GONZÁLEZ M., DEL PUERTO I., "Diffusion approximation of an array of controlled branching processes", *Methodology and Computing in Applied Probability*, vol. 14, pp. 843—861, 2012.

[GON 12b] GONZÁLEZ M., GUTIÉRREZ C., MARTÍNEZ R., "Expectation–Maximization algorithm for determining natural selection of Y–Linked genes through two–sex branching processes", *Journal of Computational Biology*, vol. 19, pp. 1015–1026, 2012.

[GON 13] GONZÁLEZ M., GUTIÉRREZ C., MARTÍNEZ R. *et al.*, "Bayesian inference for controlled branching processes through MCMC and ABC methodologies", *Revista de la Real Academia de Ciencias Exactas, Físicas y Naturales. Serie A. Matemáticas*, vol. 107, pp. 459–473, 2013.

[GON 15] GONZÁLEZ M., MINUESA C., MOTA M. *et al.*, "An inhomogeneous controlled branching process", *Lithuanian Mathematical Journal*, vol. 55, pp. 61–71, 2015.

[GON 16a] GONZÁLEZ M., GUTIÉRREZ C., MARTÍNEZ R. *et al.*, "Bayesian analysis for controlled branching processes", in DEL PUERTO I., GONZÁLEZ M., GUTIÉRREZ C. *et al.* (eds), *Branching Processes and Their Applications*, vol. 219 of *Lecture Notes in Statistics*, Springer International Publishing Switzerland, pp. 185–204, 2016.

[GON 16b] GONZÁLEZ M., MINUESA C., DEL PUERTO I., "Maximum likelihood estimation and Expectation–Maximization algorithm for controlled branching processes", *Computational Statistics & Data Analysis*, vol. 93, pp. 209–227, 2016.

[GON 17] GONZÁLEZ M., MINUESA C., DEL PUERTO I., "Minimum disparity estimation in controlled branching processes", *Electronic Journal of Statistics*, vol. 11, pp. 295–325, 2017.

[GRE 88] GREY D.R., "Supercritical branching processes with density independent catastrophes', *Mathematical Proceedings of the Cambridge Philosophical Society*, vol. 104, pp. 413–416, 1988.

[GUT 91] GUTTORP P., *Statistical Inference for Branching Processes*, Wiley, New York, 1991.

[HAC 05] HACCOU P., JAGERS P., VATUTIN V.A., *Branching Processes: Variation, Growth, and Extinction of Populations*, Cambridge University Press, Cambridge, 2005.

[HAL 49] HALDANE J.B.S., "Some statistical problems arising in genetics", *Journal of the Royal Statistical Society: Series B. Methodological*, vol. 11, pp. 1–14, 1949.

[HAL 80] HALL P., HEYDE C.C., *Martingale Limit Theory and Its Applications*, Academic Press, New York, 1980.

[HAR 63] HARRIS T.E., *Theory of Branching Processes*, Springer, New York, 1963.

[HAU 14] HAUTPHENNE S., FACKRELL M., "An EM algorithm for the model fitting of Markovian binary trees", *Computational Statistics & Data Analysis*, vol. 70, pp. 19–34, 2014.

[HEA 66] HEATHCOTE C.R., "Corrections and comments on the paper: A branching process allowing immigration", *Journal of the Royal Statistical Society: Series B. Methodological*, vol. 28, pp. 213–217, 1966.

[HEY 71] HEYDE C.C., BROWN B.M., "An invariance principle and some convergence rate results for branching processes", *Zeitschrift für Wahrscheinlichkeitstheorie und verwandte Gebiete*, vol. 20, pp. 271–278, 1971.

[HOL 84] HOLZHEIMER J., "φ-Branching processes in a random environment", *Applicationes Mathematicae*, vol. 18, pp. 351–358, 1984.

[HÖP 85] HÖPFNER R., "On some classes of population-size-dependent Galton-Watson processes", *Journal of Applied Probability*, vol. 22, pp. 25–36, 1985.

[HÖP 86] HÖPFNER R., "Some results on population–size–dependent Galton–Watson processes", *Journal of Applied Probability*, vol. 23, pp. 297–306, 1986.

[HOP 14] HOPCRAFT K.I., JAKEMAN E., RIDLEY K.D., *The Dynamics of Discrete Populations and Series of Events*, CRC Press, Boca Raton, 2014.

[IKE 89] IKEDA N., WATANABE S., *Stochastic Differential Equations and Diffusion Processes*, 2nd ed., North-Holland/Kodansha, Amsterdam, 1989.

[INA 16] INAMDAR A., KALE M., "Join estimation of offspring mean and offspring variance of controlled branching process", *Sankyā*, vol. 78, pp. 248–268, 2016.

[JAG 75] JAGERS P., *Branching Processes with Biological Applications*, Wiley, New York, 1975.

[KAL 97] KALLENBERG O., *Foundations of Modern Probability*, Springer, New York, 1997.

[KAL 00] KALE M., "Estimation of parameters in controlled branching process", *Stochastic Modelling and Applications*, vol. 3, pp. 1–15, 2000.

[KAW 71] KAWAZU K., WATANABE S., "Branching processes with immigration and related limit theorems", *Theory of Probability and Its Applications*, vol. 16, pp. 36–54, 1971.

[KEL 79] KELLY F.P., *Reversibility and Stochastic Networks*, Wiley, New York, 1979.

[KEL 87] KELLER G., KERSTING G., RÖSLER U., "On the asymptotic behaviour of discrete time stochastic growth processes", *The Annals of Probability*, vol. 15, pp. 305–343, 1987.

[KEN 51] KENDALL D.G., "Some problems in the theory of queue", *Journal of the Royal Statistical Society: Series B. Methodological*, vol. 13, pp. 151–185, 1951.

[KEN 66] KENDALL D.G., "Branching processes since 1873", *Journal of the London Mathematical Society*, vol. 41, pp. 385–406, 1966.

[KER 86] KERSTING G., "On recurrence and transience of growth models", *Journal of Applied Probability*, vol. 23, pp. 614–625, 1986.

[KER 92] KERSTING G., "Asymptotic Γ-distribution for stochastic difference equations", *Stochastic Processes and their Applications*, vol. 40, pp. 15–28, 1992.

[KHA 80] KHAN L.V., "Limit theorems for Galton–Watson processes with migration", *Siberian Mathematical Journal*, vol. 21, pp. 183–194, 1980, in Russian. English translation: vol. 21, pp. 283–292, 1980.

[KIM 15] KIMMEL M., AXELROD D.E., *Branching Processes in Biology*, 2nd ed., Springer, New York, 2015.

[KLE 84] KLEBANER F.C., "Geometric rate of growth in population–size–dependent branching processes", *Journal of Applied Probability*, vol. 21, pp. 40–49, 1984.

[KLE 85] KLEBANER F.C., "A limit theorem for population–size–dependent branching processes", *Journal of Applied Probability*, vol. 22, pp. 48–57, 1985.

[KLE 89a] KLEBANER F.C., "Geometric growth in near–supercritical population size dependent multitype Galton-Watson processes", *The Annals of Probability*, vol. 17, pp. 1466–1477, 1989.

[KLE 89b] KLEBANER F.C., "Linear growth in near-critical population–size–dependent multitype Galton–Watson processes", *Journal of Applied Probability*, vol. 26, pp. 431–445, 1989.

[KOL 38] KOLMOGOROV A.N., "Zur Lösung Einer Biologischen Aufgabe", *Izvestiya Nauchno-Issledovatelskogo Instituta Matematiki I Mechaniki pri Tomskom Gosudarstvennom Universitete*, vol. 2, pp. 1–6, 1938.

[KOS 08] KOSYGINA E., ZERNER M. P.W., "Positively and negatively excited random walks on integers, with branching processes", *Electronic Journal of Probability*, vol. 13, pp. 1952–1979, 2008.

[KUC 17] KUCHIBHOTLA A.K., BASU A., "On the asymptotics of minimum disparity estimation", *TEST*, vol. 26, pp. 481–502, 2017.

[KUE 11] KUELBS J., VIDYASHANKAR A.N., "Weak convergence results for multiple generations of a branching process", *Journal of Theoretical Probability*, vol. 24, pp. 376–396, 2011.

[KÜS 85] KÜSTER P., "Asymptotic growth of controlled Galton–Watson processes", *The Annals of Probability*, vol. 13, pp. 1157–1178, 1985.

[LAB 72] LABKOVSKII V.A., "A limit theorem for generalized random branching processes depending on the size of the population", *Theory of Probability and Its Application*, vol. 17, pp. 72–85, 1972.

[LAM 67] LAMPERTI J., "Continuous state branching processes", *Bulletin of the American Mathematical Society*, vol. 73, pp. 382–386, 1967.

[LEV 68] LEVINA L.V., LEONTOVICH A.M., PYATETSKII-SHAPIRO I.I., "A controllable branching process", *Problems of Information Transmission*, vol. 4, pp. 72–82, 1968.

[LIN 94] LINDSAY B.G., "Efficiency versus robustness: The case for minimum Hellinger distance and related methods", *The Annals of Statistics*, vol. 22, pp. 1081–1114, 1994.

[LOÈ 17] LOÈVE M., *Probability Theory*, 3rd ed., Dover Publications, Mineola, 2017.

[LYO 16] LYONS R., PERES Y., *Probabilities on Trees and Networks*, Cambridge University Press, Cambridge, 2016.

[MAR 09] MARTÍNEZ R., MOTA M., DEL PUERTO I., "On asymptotic posterior normality for controlled branching processes with random control function", *Statistics*, vol. 43, pp. 367–378, 2009.

[MAY 05] MAYSTER P., "Alternating branching processes", *Journal of Applied Probability*, vol. 42, pp. 1095–1108, 2005.

[MCL 08] MCLACHLAN G., KRISHNAN T., *The EM Algorithm and Extensions*, Wiley, New York, 2008.

[MIT 84a] MITOV K.V., VATUTIN V.A., YANEV N.M., "Continuous–time branching processes with decreasing state-dependent immigration", *Advances in Applied Probability*, vol. 16, pp. 697–714, 1984.

[MIT 84b] MITOV K.V., VATUTIN V.A., YANEV N.M., "Critical Galton-Watson processes with decreasing, state dependent immigration" (in Russian), *Serdica Mathematical Journal*, vol. 10, pp. 412–424, 1984,

[MIT 01] MITOV K.V., YANEV N.M., "Limit theorems for alternating renewal processes in the infinite mean case", *Advances in Applied Probability*, vol. 33, pp. 896–911, 2001.

[MIT 02] MITOV K.V., YANEV N.M., "Critical Bellman–Harris branching processes with infinite variance allowing state dependent immigration", *Stochastic Models*, vol. 18, pp. 281–300, 2002.

[MIT 09] MITOV K.V., YANEV N.M., "Branching stochastic processes: regulation, regeneration, estimation, applications", *Pliska Studia Mathematica Bulgarica*, vol. 19, pp. 5–58, 2009.

[MOL 98] MOLINA M., GONZÁLEZ M., MOTA M., "Some theoretical results about superadditive controlled Galton-Watson branching processes.", in HUSKOVA M., LACHOUT P., VISEK J.A. (eds), *Proceedings of Prague Stochastics'98*, vol. 1, Union of Czech Mathematicians and Physicists, Prague, Czech Republic, pp. 413–418, 1998.

[MOL 10] MOLINA M., "Two-sex branching process literature", in GONZÁLEZ M., DEL PUERTO I., MARTÍNEZ R. *et al.* (eds), *Workshop on Branching Processes and Their Applications*, vol. 197 in *Lecture Notes in Statistics*, Springer, Berlin, pp. 279–293, 2010.

[NAG 80] NAGAEV S.V., HAN L.V., "Limit theorems for critical Galton-Watson branching process with migration", *Theory of Probability and Its Applications*, vol. 25, pp. 523–534, 1980, in Russian. English translation: vol. 25, pp. 514–525, 1981.

[NAK 90] NAKAGAWA T., "On the geometric growth of controlled branching processes in a random environment", *Bulletin of General Education, Dokkyo University School of Medicine*, vol. 13, pp. 53–60, 1990.

[NAK 94] NAKAGAWA T., "The L^α $(1 < \alpha \leq 2)$ convergence of a controlled branching process in a random environment", *Bulletin of General Education, Dokkyo University School of Medicine*, vol. 17, pp. 17–24, 1994.

[PAK 71a] PAKES A.G., "A branching process with a state dependent immigration component", *Advances in Applied Probability*, vol. 3, pp. 301–314, 1971.

[PAK 71b] PAKES A.G., "Branching processes with immigration", *Journal of Applied Probability*, vol. 8, pp. 32–42, 1971.

[PAK 71c] PAKES A.G., "On the critical Galton–Watson process with immigration", *Journal of the Australian Mathematical Society*, vol. 12, pp. 476–482, 1971.

[PAK 74] PAKES A.G., "On supercritical Galton–Watson processes allowing immigration", *Journal of Applied Probability*, vol. 11, pp. 814–817, 1974.

[PAK 75] PAKES A.G., "Some results for non–supercritical Galton–Watson processes with immigration", *Mathamatical Biosciences*, vol. 24, pp. 71–92, 1975.

[PAK 79] PAKES A.G., "Limit theorems for the simple branching process allowing immigration, I. The case of finite offspring mean", *Advances in Applied Probability*, vol. 11, pp. 31–62, 1979.

[PAK 84] PAKES A.G., "The Galton–Watson process with killing", *Mathematical Biosciences*, vol. 69, pp. 171–188, 1984.

[PAK 06] PAKES A.G., "Immigration–Emigration Processes", in KOTZ S., READ C., BALAKRISHNAN N. *et al.* (eds), *Encyclopedia of Statistical Sciences*, Wiley, New York, pp. 3344–3347, 2006.

[PAR 06] PARDO L., *Statistical Inference Based on Divergence Measures.*, Chapman and Hall/CRC, Boca Raton, 2006.

[PUE 07] DEL PUERTO I., YANEV N.M., "Stationary distributions for branching processes with multitype random control functions", *Journal of Applied Statistical Science*, vol. 16, pp. 91–101, 2007.

[RAH 86] RAHIMOV I., "Critical branching processes with infinite variance and decreasing immigration", *Theory of Probability and Its Applications*, vol. 31, pp. 98–110, 1986, in Russian. English translation: vol. 31, no. 1, pp. 88–100, 1987.

[RAH 87] RAHIMOV I., KAVERIN S.V., "On a method of proof of limit theorems for the branching process with state-dependent immigration" (in Russian), *Probabilistic Models and Mathematical Statistics*, FAN, Tashkent, pp. 61–76, 1987.

[RAH 95] RAHIMOV I., *Random Sums and Branching Stochastic Processes*, Springer, New York, 1995.

[RAH 07] RAHIMOV I., "Asymptotic behavior of a controlled branching process with continuous state space", *Stochastic Analysis and Applications*, vol. 25, pp. 337–352, 2007.

[RAH 17] RAHIMOV I., "Statistical inference for partially observed branching processes with immigration", *Journal of Applied Probability*,, vol. 54, pp. 82–95, 2017.

[SAG 17] SAGITOV S., "Tail generating functions for extendable branching processes", *Stochastic Processes and their Applications*, vol. 127, pp. 1649–1675, 2017.

[SEN 68] SENETA E., "The stationary distribution of a branching process allowing immigration: a remark on the critical case", *Journal of the Royal Statistical Society: Series B. Methodological*, vol. 30, pp. 176–179, 1968.

[SEN 69] SENETA E., "Functional equations and the Galton-Watson process", *Advances in Applied Probability*, vol. 1, pp. 1–42, 1969.

[SEN 70] SENETA E., "A note on the supercritical Galton-Watson process with immigration", *Mathematical Biosciences*, vol. 6, pp. 305–311, 1970.

[SEV 57] SEVASTYANOV B.A., "Limit theorems for branching processes of special form", *Theory of Probability and Its Applications*, vol. 2, pp. 339–348, 1957, in Russian. English translation: vol. 2, pp. 321–331, 1957.

[SEV 71] SEVASTYANOV B.A., *Branching Processes* (in Russian), Nauka, Moscow, 1971.

[SEV 74] SEVASTYANOV B.A., ZUBKOV A.M., "Controlled branching processes", *Theory of Probability and Its Applications*, vol. 19, pp. 15–25, 1974.

[SEV 09] SEVASTYANOV B.A., "Interview", in DEMIDOVICH V.B. (ed.), *Recollections of the Alumni of Mathematics Department at Moscow University II* (in Russian), Moscow University, Moscow, 2009.

[SHE 03] SHETE S., SRIRAM T.N., "A note on estimation in multitype supercritical branching processes with immigration", *Sankhyā*, vol. 65, pp. 107–121, 2003.

[SLA 68] SLACK R.S., "A branching process with mean one and possibly infinite variance", *Zeitschrift für Wahrscheinlichkeitstheorie und verwandte Gebiete*, vol. 9, pp. 139–145, 1968.

[SRI 94] SRIRAM T.N., "Invalidity of bootstrap for critical branching processes with immigration", *The Annals of Statistics*, vol. 22, pp. 1013–1023, 1994.

[SRI 00] SRIRAM T.N., VIDYASHANKAR A.N., "Minimum Hellinger distance estimation for supercritical Galton–Watson processes", *Statistics and Probability Letters*, vol. 50, pp. 331–342, 2000.

[SRI 07] SRIRAM T.N., BHATTACHARYA A., GONZÁLEZ M. *et al.*, "Estimation of the offspring mean in a controlled branching process with a random control function", *Stochastic Processes and their Applications*, vol. 117, pp. 928–946, 2007.

[STO 04] STOIMENOVA V., ATANASOV D., YANEV N.M., "Robust estimation and simulation of branching processes", *Comptes rendus de l'Académie Bulgare des Sciences*, vol. 57, pp. 19–22, 2004.

[TIE 94] TIERNEY L., "Markov chains for exploring posterior distributions", *The Annals of Statistics*, vol. 22, pp. 1701–1762, 1994.

[ULA 90] ULAM S.M., *Analogies Between Analogies. The Mathematical Reports of S.M. Ulam and His Los Alamos Collaborators*, University of California Press, Berkeley, 1990.

[VAT 77] VATUTIN V.A., "A critical Galton–Watson branching process with emigration", *Theory of Probability and Its Applications*, vol. 22, pp. 482–497, 1977, in Russian. English translation: vol. 22, pp. 465–481, 1977.

[VAT 79] VATUTIN V.A., "New limit theorem for the critical Bellman–Harris branching process", *Mathematics of the USSR - Sbornik*, vol. 109, pp. 440–452, 1979, in Russian. English translation: vol. 37, pp. 411–423, 1980.

[VAT 93] VATUTIN V.A., ZUBKOV A.M., "Branching Processes II", *Journal of Soviet Mathematics*, vol. 67, pp. 3406–3485, 1993.

[VAT 08] VATUTIN V.A., *Branching Processes and Their Applications* (in Russian), *Steklov Institute of Mathematics*, Russian Academy of Sciences, Moskva, 2008.

[VEE 08] VEEN A., SCHOENBERG F.P., "Estimation of space–time branching process models in seismology using an EM–Type Algorithm", *Journal of the American Statistical Association*, vol. 103, pp. 614–624, 2008.

[WEI 90] WEI C.Z., WINNICKI J., "Estimation of the mean in the branching process with immigration", *The Annals of Statistics*, vol. 18, pp. 1757–1773, 1990.

[WIL 91] WILLIAMS D., *Probability with Martingales*, Cambridge University Press, Cambridge, 1991.

[WOL 89] WOLFF R.W., *Stochastic Modeling and the Theory of Queues*, Prentice Hall, Englewood Cliffs, 1989.

[YAG 47] YAGLOM A.M., "Certain limit theorems in the theory of branching stochastic processes" (in Russian), *Doklady Akademii Nauk*, vol. 56, pp. 795–798, 1947.

[YAN 75] YANEV N.M., "Conditions for degeneracy of φ- branching processes with random φ", *Theory of Probability and Its Applications*, vol. 20, pp. 433–440, 1975.

[YAN 77] YANEV N.M., "Controlled branching processes in random environment", *Mathematica Balkanica (N.S.)*, vol. 7, pp. 137–156, 1977.

[YAN 80] YANEV N.M., MITOV K.V., "Controlled branching processes. The case of random migration", *Comptes rendus de l'Académie bulgare des Sciences*, vol. 33, pp. 473–475, 1980.

[YAN 82] YANEV N.M., MITOV K.V., "Limit theorems for controlled branching processes with nonhomogeneous migration", *Comptes rendus de l'Académie bulgare des Sciences*, vol. 35, pp. 299–301, 1982.

[YAN 83] YANEV N.M., MITOV K.V., "The life–periods of critical branching processes with random migration", *Theory of Probability and Its Applications*, vol. 28, pp. 458–467, 1983.

[YAN 84a] YANEV N.M., MITOV K.V., "Controlled branching processes with non–homogeneous migration", *Pliska Studia Mathematica Bulgarica*, vol. 7, pp. 90–96, 1984.

[YAN 84b] YANEV N.M., MITOV K.V., "Limit theorems for controlled branching processes with decreasing emigration", *Pliska Studia Mathematica Bulgarica*, vol. 7, pp. 83–89, 1984, in Russian.

[YAN 84c] YANEV N.M., MITOV K.V., "Subcritical branching processes with migration (in Russian)", *Pliska Studia Mathematica Bulgarica*, vol. 7, pp. 75–82, 1984, in Russian.

[YAN 85a] YANEV N.M., MITOV K.V., "A critical branching process with decreasing migration", *Serdica Mathematical Journal*, vol. 11, pp. 240–244, 1985.

[YAN 85b] YANEV N.M., MITOV K.V., "Critical branching processes with nonhomogeneous migration", *The Annals of Probability*, vol. 13, pp. 923–933, 1985.

[YAN 89] YANEV G.P., YANEV N.M., "Conditions for extinction of controlled branching processes", *Mathematics and Education in Mathematics*, The Union of Bulgarian Mathematicians, Sofia, pp. 550–555, 1989.

[YAN 90] YANEV G.P., YANEV N.M., "Extinction of controlled branching processes in random environments", *Mathematical Balkanica (N.S.)*, vol. 4, pp. 368–380, 1990.

[YAN 95] YANEV G.P., YANEV N.M., "Critical branching process with random migration", in HEYDE C.C. (ed.), *Branching Processes. Proceedings of the First World Congress*, vol. 99 of *'Lecture Notes in Statistics'*, Springer, New York, pp. 36–46, 1995.

[YAN 96] YANEV G.P., YANEV N.M., "Branching processes with two types emigration and state-dependent immigration", in HEYDE C.C., PROHOROV Y.V., PYKE R. *et al.* (eds), *Athens Conference on Applied Probability and Times Series*, vol. 114 of *'Lecture Notes in Statistics'*, Springer, New York, pp. 216–228, 1996.

[YAN 97] YANEV G.P., YANEV N.M., "Limit theorems for branching processes with random migration stopped at zero", in ATHREYA K.B., JAGERS P. (eds), *Classical and Modern Branching Processes*, vol. 84 of *The IMA Volumes in Mathematics and its Application*, Springer, New York, pp. 323–336, 1997.

[YAN 00] YANEV G.P., YANEV N.M., "Limit theorems for branching processes with random migration components", *Pliska Studia Mathematica Bulgarica*, vol. 13, pp. 199–204, 2000.

[YAN 03] YANEV G.P., MITOV K.V., YANEV N.M., "Critical branching processes regenerative processes with migration", *Journal of Applied Statistical Science*, vol. 12, pp. 41–54, 2003.

[YAN 04] YANEV G.P., YANEV N.M., "A critical branching process with stationary-limiting distribution", *Stochastic Analysis and Applications*, vol. 22, pp. 721–738, 2004.

[YAN 08] YANEV N.M., "Statistical inference for branching processes", in AHSANULLAH M., YANEV G.P., (eds), *Records and Branching Processes*, Nova Science, New York, 2008.

[YAN 10] YANG S., "A fluctuation limit theorem of branching processes and its applications", *Statistical Sinica*, vol. 20, pp. 1273–1289, 2010.

[YAN 15] YANEV G.P., "Critical controlled branching processes and their relatives", *Pliska Studia Mathematica Bulgarica*, vol. 24, pp. 111–130, 2015.

[ZUB 72] ZUBKOV A.M., “Life–periods of a branching process with immigration”, *Theory of Probability and Its Applications*, vol. 17, pp. 179–188, 1972, in Russian. English translation: vol. 17, pp. 174–183, 1972.

[ZUB 74] ZUBKOV A.M., “Analogies between Galton–Watson processes and φ-branching processes”, *Theory of Probability and Its Applications*, vol. 19, pp. 319–339, 1974.

Index

A

B

C

D

E, F

G

I, K

L, M

O, P

S, T

Other titles from

iSTE

in

Mathematics and Statistics

2017

CARONI Chysseis
First Hitting Time Regression Models: Lifetime Data Analysis Based on Underlying Stochastic Processes
(Mathematical Models and Methods in Reliability Set – Volume 4)

CELANT Giorgio, BRONIATOWSKI Michel
Interpolation and Extrapolation Optimal Designs 2: Finite Dimensional General Models

CONSOLE Rodolfo, MURRU Maura, FALCONE Giuseppe
Earthquake Occurrence: Short- and Long-term Models and their Validation
(Statistical Methods for Earthquakes Set – Volume 1)

D'AMICO Guglielmo, DI BIASE Giuseppe, JANSSEN Jacques, MANCA Raimondo
Semi-Markov Migration Models for Credit Risk
(Stochastic Models for Insurance Set – Volume 1)

HARLAMOV Boris
Stochastic Analysis of Risk and Management
(Stochastic Models in Survival Analysis and Reliability Set – Volume 2)

KERSTING Götz, VATUTIN Vladimir
Discrete Time Branching Processes in Random Environment
(Branching Processes, Branching Random Walks and Branching Particle Set – Volume 1)

MISHURA YULIYA, SHEVCHENKO Georgiy
Theory and Statistical Applications of Stochastic Processes

NIKULIN Mikhail, CHIMITOVA Ekaterina
Chi-squared Goodness-of-fit Tests for Censored Data
(Stochastic Models in Survival Analysis and Reliability Set – Volume 3)

SIMON Jacques
Banach, Fréchet, Hilbert and Neumann Spaces
(Analysis for PDEs Set – Volume 1)

2016

CELANT Giorgio, BRONIATOWSKI Michel
Interpolation and Extrapolation Optimal Designs 1: Polynomial Regression and Approximation Theory

CHIASSERINI Carla Fabiana, GRIBAUDO Marco, MANINI Daniele
Analytical Modeling of Wireless Communication Systems
(Stochastic Models in Computer Science and Telecommunication Networks Set – Volume 1)

GOUDON Thierry
Mathematics for Modeling and Scientific Computing

KAHLE Waltraud, MERCIER Sophie, PAROISSIN Christian
Degradation Processes in Reliability
(Mathematial Models and Methods in Reliability Set – Volume 3)

KERN Michel
Numerical Methods for Inverse Problems

RYKOV Vladimir
Reliability of Engineering Systems and Technological Risks
(Stochastic Models in Survival Analysis and Reliability Set – Volume 1)

2015

DE SAPORTA Benoîte, DUFOUR François, ZHANG Huilong
Numerical Methods for Simulation and Optimization of Piecewise Deterministic Markov Processes

DEVOLDER Pierre, JANSSEN Jacques, MANCA Raimondo
Basic Stochastic Processes

LE GAT Yves
Recurrent Event Modeling Based on the Yule Process
(Mathematical Models and Methods in Reliability Set – Volume 2)

2014

COOKE Roger M., NIEBOER Daan, MISIEWICZ Jolanta
Fat-tailed Distributions: Data, Diagnostics and Dependence
(Mathematical Models and Methods in Reliability Set – Volume 1)

MACKEVIČIUS Vigirdas
Integral and Measure: From Rather Simple to Rather Complex

PASCHOS Vangelis Th
Combinatorial Optimization 3-volume series – 2nd edition
Concepts of Combinatorial Optimization / Concepts and Fundamentals – volume 1
Paradigms of Combinatorial Optimization – volume 2
Applications of Combinatorial Optimization – volume 3

2013

COUALLIER Vincent, GERVILLE-RÉACHE Léo, HUBER Catherine, LIMNIOS Nikolaos, MESBAH Mounir
Statistical Models and Methods for Reliability and Survival Analysis

JANSSEN Jacques, MANCA Oronzio, MANCA Raimondo
Applied Diffusion Processes from Engineering to Finance

SERICOLA Bruno
Markov Chains: Theory, Algorithms and Applications

2012

BOSQ Denis
Mathematical Statistics and Stochastic Processes

CHRISTENSEN Karl Bang, KREINER Svend, MESBAH Mounir
Rasch Models in Health

DEVOLDER Pierre, JANSSEN Jacques, MANCA Raimondo
Stochastic Methods for Pension Funds

2011

MACKEVIČIUS Vigirdas
Introduction to Stochastic Analysis: Integrals and Differential Equations

MAHJOUB Ridha
Recent Progress in Combinatorial Optimization – ISCO2010

RAYNAUD Hervé, ARROW Kenneth
Managerial Logic

2010

BAGDONAVIČIUS Vilijandas, KRUOPIS Julius, NIKULIN Mikhail
Nonparametric Tests for Censored Data

BAGDONAVIČIUS Vilijandas, KRUOPIS Julius, NIKULIN Mikhail
Nonparametric Tests for Complete Data

IOSIFESCU Marius *et al.*
Introduction to Stochastic Models

VASSILIOU PCG
Discrete-time Asset Pricing Models in Applied Stochastic Finance

2008

ANISIMOV Vladimir
Switching Processes in Queuing Models

FICHE Georges, HÉBUTERNE Gérard
Mathematics for Engineers

HUBER Catherine, LIMNIOS Nikolaos *et al.*
Mathematical Methods in Survival Analysis, Reliability and Quality of Life

JANSSEN Jacques, MANCA Raimondo, VOLPE Ernesto
Mathematical Finance

2007

HARLAMOV Boris
Continuous Semi-Markov Processes

2006

CLERC Maurice
Particle Swarm Optimization

Printed and bound by CPI Group (UK) Ltd, Croydon, CR0 4YY
16/09/2026
14963181-0003